APPLIED
FOODSERVICE SANITATION

APPLIED
FOODSERVICE SANITATION

A NIFI Textbook

Published by the
National Institute for the Foodservice Industry

**Developed in Collaboration with the
National Sanitation Foundation**

© 1974 by the NATIONAL INSTITUTE
FOR THE FOODSERVICE INDUSTRY

120 S. Riverside Plaza, Chicago, Ill. 60606
All rights reserved. Printed in the United States of America

LIBRARY OF CONGRESS CATALOGING IN PUBLICATION DATA
National Institute for the Foodservice Industry.
    Applied foodservice sanitation.
    Bibliography: p.
    1. Food service—Sanitation. I. National Sanitation
Foundation, Ann Arbor, Mich. II. Title.
TX943.N37   1974   614.7'91   74-3284

*This book is inscribed
to the memory of*

LEWIS ATHANAS

*brother of Anthony Athanas
of Anthony's Pier 4,
Boston, Massachusetts,
whose generous grant to
the Institute in large part
made its publication
possible.*

Foreword

Over the past quarter century, food service has changed significantly. No longer are we limited in our menu by seasonal foods or in the variety of foods and style of presentation. With all these changes—with food supplies being shipped internationally—it is most important that the food-service manager and the food-service employee understand the basic facts of food sanitation and how to prevent foodborne illness.

In the past, many health agencies have attempted to train "food handlers" through infrequently scheduled classes, usually with little or no food-service management attendance or support. More often than not, such sessions were ineffective. This too has changed. The production of this text by the National Institute for the Foodservice Industry is an indication of the "forward looking" concept. The food-service industry is assuming its rightful role of training its employees. It is to be commended for this effort.

Protection of the health of the consumer is our greatest challenge. Through understanding and implementing the principles of food protection, the industry and the regulatory agency together can meet that challenge.

William F. Bower, Acting Director
Division of Food Service
Food and Drug Administration
U.S. Public Health Service

Contents

A Message from the Institute

The National Institute for the Foodservice Industry is pleased to present *Applied Foodservice Sanitation,* a NIFI book for foodservice managers, supervisors and those aspiring to management jobs in the industry. It is a book to be read and studied. Its purpose is to illuminate a very important subject, but one that is not always well understood. Contamination hazards threaten our food supply at every twist and turn of the long road from farm to table. In dealing with these hazards the problem is to convince everyone concerned that rather simple techniques are available to us for avoiding them.

The responsibility of management to protect the public from foodborne illness is fundamental. For the foodservice manager in particular it is a responsibility that cannot be transferred. To be sure, he must train his people to handle food safely, but his task does not end there. To ensure that the health of the consumer is fully safeguarded, he will inevitably have to stand behind the performance of every foodhandler on his staff.

Applied Foodservice Sanitation has been developed to prepare the manager for this challenging task. Scientifically and technologically, we are on firm ground. The epidemiological problem is well documented. Public awareness of the dangers inherent in unsafe foodservice operations has been sharpened, sometimes in spectacular ways, by recent attention in the public information media.

We trust the present volume will deserve a place on every manager's bookshelf. In any case it is not merely another book on an important subject. It is the central text for a NIFI management course designed to reach the greatest possible number of people entrusted with serving food to the general public. As with other management courses in the NIFI series, the sanitation course is arranged for administration in a variety of ways. Individuals may undertake it as a home study project in direct correspondence with the Institute. Foodservice personnel may take the course in a format tailored for in-service group training programs sponsored by foodservice companies, trade associations and other industry organizations. The text and related materials will, of course, be available to college and community college students of hotel, restaurant and institutional management as part of the regular academic curriculum.

All who complete the course—through home study, in industry groups or in the college classroom—will be eligible for a NIFI certificate upon satisfactory completion of a certification examination administered under Institute auspices.

A few words about the National Institute. NIFI is our industry's not-for-profit educational foundation, created by restaurateurs and other food-service executives, and governed by a board of trustees representing all sectors of the industry and associated academic and commercial institutions. The Institute is dedicated to helping career-minded men and women take advantage of the expanding opportunities in America's dining-out industry. In so doing, it renders a vital service to the industry itself, the largest employer in the country in terms of the number of people employed. Food-service will grow and prosper only as the quality of its trained personnel grows and prospers.

The most critical and longstanding problem of the foodservice industry, today and in the foreseeable future, is obtaining, training and retaining professional managers and supervisors to keep pace with the burgeoning public demand for meals away from home. Not less than 25,000 such professionals, qualified to provide the highest standards of service, must be found each year in the next decade. Nothing else, we recognize, will satisfy our guests.

The Institute's answer is education—a career-opportunity program enabling foodservice employees and students to increase their knowledge and qualifications and move quickly and effectively into front-line management positions. Management education and training keyed to industry needs is a major part of our mission. NIFI certification of professional managers who meet prescribed standards of education, experience and on-the-job performance is another.

Due regard for sanitary standards with respect to the food he serves and the service he provides is essential for any foodservice manager who calls himself a professional. The qualified manager knows that:

Micro-organisms are everywhere—and they belong there.
Some are good and some are bad.
They can be controlled.
Centralized processing of pre-cooked, "convenience" foods magnifies the sanitation risks involved.
Despite advances in food technology and quality control, foodborne illness is still with us.
We cannot leave sanitation to the sanitarians.
Protecting the public is not only the law: it's good business.

Learning to "manage the microbe" can be an interesting as well as rewarding pursuit . . . and that's where this book comes in.

Chester G. Hall Jr., Ph.D.
Executive Vice President
National Institute
   for the Foodservice Industry

Editor's Preface

This textbook is written for the reader and student. It is designed to encourage an enlightened approach to problems that are commonplace but nonetheless in some degree technical. Although the writers treat the subject operationally, they seek withal to give the foodservice manager and staff an understanding of the fundamental scientific principles underlying good sanitation practice.

Development of the text and related course materials has been a joint undertaking of the National Institute for the Foodservice Industry and the National Sanitation Foundation. The basic content is largely attributable to the work of D. L. Lancaster, of the Foundation's education department, who assembled the material and prepared the original manuscript. Valuable guidance from educators and government authorities in the field is gratefully acknowledged. We are especially indebted to the Division of Food Service, U.S. Food and Drug Administration, for its forthright critique and recommendations, and to the National Environmental Health Association for a review of the book from the standpoint of the professional sanitarian. Our very special thanks go to Vernon E. Cordell, Director of Public Health and Safety, National Restaurant Association, who has been a steadfast counsellor from the beginning.

HOW TO READ THIS BOOK. Chapter by chapter and in its general scheme the book follows a time-honored instructional method: "Tell 'em what you're going to tell 'em; tell 'em; and then tell 'em what you told 'em." Chapter 1 is thus an overview of the subject, and Chapter 14 recapitulates its main themes in a summary statement for the foodservice manager. The reader is advised to begin with a preview of both the first and final chapters, and to return to these pages from time to time to keep his studies on the track.

A note on terminology: In the interest of readability we have sought to avoid technical terms and, when they were necessary, to introduce them in a logical way as the subject matter developed. Words likely to be unfamiliar will also be defined in the glossary at the back of the book.

Floyd J. Greene

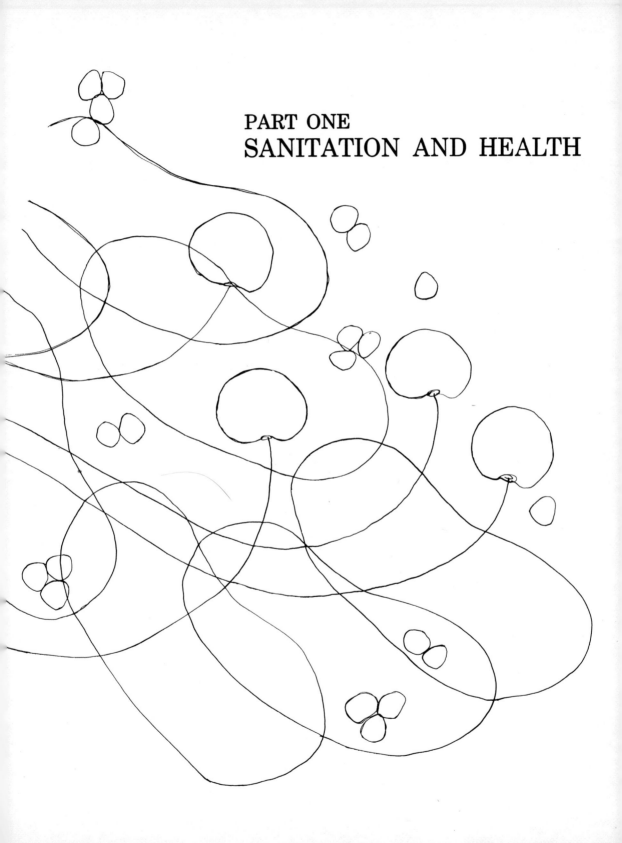

PART ONE
SANITATION AND HEALTH

CHAPTER 1

The Problem of Safe Food

The traditional restaurant operator hardly thinks of himself as a technical expert on food. Nor does the family cook. But, in catering to one of our most fundamental needs, both of them literally hold our health in their hands. Next to the air we breathe, what could be more important than the food we eat and the water we drink? (Granting, of course, the gift of life, the good earth, and other blessings of a bounteous Providence.)

Whether it's Mom or the professional cook, and consciously or not, the person who feeds us presides over a process that runs the gamut of the chemical and biological sciences, to say nothing of the arts.

The typical restaurateur probably regards his role primarily as that of a businessman providing a vital service. He might fancy himself, or his chef, to be a culinary artist, but most likely he would balk at being called a nutritionist, and certainly would not welcome the title of sanitarian. These last two concepts are nevertheless looming up more and more importantly in his operations nowadays. Mass production of food and the growing trend toward mass preparation

and serving, of pre-processed foods in particular, are partly responsible for the change.

There will always be room for romance in cooking and adventure in good eating, we trust, but the modern foodservice specialist must surely forsake the pre-scientific, folklore approach to nutrition and health. Indeed, we have already witnessed significant moves in this direction. In a generation or less, home economics has reached beyond the classroom, as modern homemakers have become increasingly aware of the principles of good nutrition. Likewise, on the commercial front, enrollment in management courses in hotel, restaurant and institutional feeding continues on the rise, as the foodservice industry quickens its march into professionalism.

Research in food chemistry and the more creative aspects of food technology appears to be in good hands. So, too, does its counterpart in the protective area: the fight against foodborne illness. At least the call to arms has been sounded. In the colorful words of a leading foodservice sanitarian of our acquaintance: "The battle lines are drawn at the bridgehead guarding the strategic Alimentary Canal. It's a game of a-l-i-m-e-n-t versus a-i-l-m-e-n-t, and our job is to see that the I's and the L's don't get crossed."

With a nod to the nutritionists, then, and never forgetting our duties in that sphere, we turn to the zone of defense. Our purpose is to examine the problems faced by the foodservice manager in safeguarding the health of the public. His motives are a complex of ethical, legal, and economic considerations, but he has a clear-cut single objective: *To protect people against illness from contamination of food by harmful organisms and their toxins, and by other poisonous materials.*

In meeting this responsibility, the foodservice manager is committed primarily to one or both of two courses of action:

- **Keep food free of bacterial contaminants in the first place.**

- **Prevent the growth of bacteria that may invade food during storage, preparation, and service.**

How he does this is, of course, the principal thrust of this book.

Ask a sanitation-conscious manager how he goes after the problem, and you may get a surprisingly brief reply, like: "Trained people and plenty of hot water." A typical food sanitarian lecturing to a group of foodhandlers may sum up his remarks with the advice: "Serve it hot or keep it cold, wash your hands, and don't sneeze!"

Reduced to the simplest terms, protecting people from illness due to contaminated food does appear to be a pretty straightforward matter. It is. The

stumbling block comes in awakening people to the existence of the problem, and getting them to act on it — by habit.

In this section we will highlight some of the guideposts along the road to a safe foodservice, but before proceeding let's examine . . .

HOW WE FEEL ABOUT SANITATION

As you embark on this seemingly unglamorous phase of your education in foodservice management, nothing should take precedence over your attitude toward the subject. This attitude must of course be based upon knowledge of the industry, an appreciation of the dimensions of the foodborne illness problem, and your professional and legal position in relation to it.

The Stakes are High

Why are we concerned about food contamination and its effect on people's health? Without resorting to an elaborate numbers game, let's pause to digest a few statistics. The foodservice industry is now rated as the third largest retail industry in terms of the dollar value of its national product, $45,000,000,000 a year. It is generally regarded as one of the fastest-growing industries in the United States, and had a total employment of 3,700,000 persons in 1973.

In the 1970's, Americans will be eating one-fourth to one-third of their meals away from the family table—in restaurants, school and company cafeterias, hospitals, and other foodservice operations. For those who provide the meals — the restaurateurs and other foodservice operators — that's the *good* news.

The *bad* news is that in 1972 an estimated 6,000,000 people contracted foodborne illness eating away from home. This number probably reflects an increase in the absolute total over past years but not necessarily an upward trend on a percentage basis. Are diners being made ill at a higher rate than previously? We don't know for sure. It may only be that food poisoning incidents are now being reported more completely.

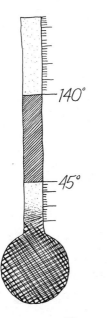

Fig. 1 The temperature "danger zone"

Although such a figure is disturbing any way you look at it, it is not to be taken as an outright indication of imminent danger to the general public. A more reasonable posture is one of precaution. It is the *potential* hazard which we should have mainly in mind, with an eye on the profusion of new eating places dotting the landscape and their insatiable demand for trained people to operate them. Many of these new places are units of large national chains, and some are also mass-feeding industrial foodservices. In either case, the possibility of affecting large numbers of people is increased by at least two factors: the sheer mass of food being processed, shipped and served, and the nature of many new, precooked food products coming into the market. The last-cited factor is primarily one of time—the extended time during which it is possible for accidents and mishandling to occur, and especially the added occasions for excessive bacterial growth due to unsafe holding (not hot enough) and unsafe keeping (not cold enough) temperatures. (Fig. 1) The critical importance of temperature control in combatting pathogenic (disease-causing) micro-organisms is treated fully in a later chapter.

There is a Law

The most powerful incentive to action for the loyal, trained worker is an order from the boss. Foodservice people are not a highly regimented group in our society, but legal incentives are certainly not lacking. State laws and municipal ordinances are more than adequate in outlawing unsafe foodhandling practices on the local scene; federal regulations cover food produced and processed for interstate commerce. (Indeed, any food manufacturer or restaurant operator will likely sound off about the "burden" of public health inspections.) But this is not the final answer. It ultimately comes down to the individual foodhandler with an infected finger. If his attitude is right and he is trained, he will instinctively avoid unsafe practices. That's the real answer.

Speaking of attitudes, it has often been said that the thing we fear most is the unknown. Conversely, we like the thing we know best. It would be unfair to leave our discussion of motives without rebutting the suggestion that sanitation is an unglamorous subject and an unexciting part of the foodservice manager's job. When we truly know something about it, we see it another way. It becomes a manageable task, even an interesting one. For many students of microbiology it is clearly a fascinating subject leading to a lifetime career. This brings us to the micro-world and what it means to the foodservice sanitarian. It means a very great deal, you can be sure, and not only to one who is already "sold." An appreciation of it will help to "sell" you.

FIRST YOU SEE THE INVISIBLE

Do yourself a favor. Think about germs.

"It's unnatural," you say. "It's unpleasant. Why should I?"

Well, let's see. We don't have to think about breathing. We don't have to think about walking. Or dropping a hot potato. So far, you're right. We don't have to think consciously about a lot of things we do — and for that we can thank Heaven. We would be in a pretty pickle if our "reflexes" didn't govern innumerable body functions—vital functions we couldn't voluntarily control even if we wanted to. We are hot: we perspire and cool off. We cut our hand: blood rushes antibodies to the area and healing starts immediately. Food enters our mouth: without a thought from us, saliva springs

forth and other digestive juices begin to flow. If myriad tiny, invisible "hands" were not always at work, unbidden by us, there could be no living thing. Life would never have evolved.

So, why bother? The answer is that, where disease-causing germs are concerned, including pathogens that invade food, we are not completely helpless. Although we couldn't begin to eliminate them entirely, we can keep them in their place. In that sense, sanitation may be seen as another example of "rearranging the dirt." But that we can do. That we must do.

Quite obviously, the first priority is to awaken ourselves to the supreme fact that we are surrounded by unseen forms of life on every hand. Succeeding chapters will deal in some detail with the world of micro-organisms, food contamination, and foodborne illness. At this point it is enough for us to recognize that:

- **Micro-organisms are everywhere.**
- **They *mean* something—good and bad—to all of us.**
- **They can be managed.**

FOOD CONTAMINATION DEFINED

There's a curious thing about proteins, vitamins, enzymes, bacteria, yeast, viruses—all the catalysts and other microscopic agents in our digestive system. These very same mechanisms not only break down nutrients and convert them to tissue and energy; they are also instrumental in another kind of biochemical conversion—the kind that wrecks food and poisons the consumer.

It all depends on what's breaking down. And when. And where. The nutritionist is interested in what makes food food, and how much of each kind we need. We are interested in keeping safe food safe.

But what, specifically, do we mean by food contamination as it relates to the foodservice operation?

Many people are inclined to equate food contamination and food spoilage. These two phenomena represent similar biochemical events, but it should be noted that the two are not identical. The kind of contamination we are primarily concerned with is that which introduces disease agents into the food. But an agent which we regard as a spoiler in one instance may be useful in another. The bacterial ferment that sours milk but is used in making cheese is a familiar example.

Our main attention will be given to "potentially hazardous" foods. As defined by the U.S. Public Health Service, these are foods that are capable of supporting rapid and progressive growth of infectious or toxicogenic (poison-producing) micro-organisms. For the most part, we are only indirectly concerned with "perishable" and "nonperishable" foods referred to in the official literature, since perishability and resultant spoilage are only incidentally related to the contamination which causes foodborne illness. In most cases food which has spoiled or deteriorated to the point of losing color, texture, flavor and desirable aroma, will not cause illness. On the other hand, food contaminated with harmful organisms and their toxins does not always betray this condition in its appearance or taste.

It is true, however, that conditions favorable to the growth of spoilage organisms are often favorable to the growth of infectious or toxic organisms, and that many of the procedures which protect food from spoilage also serve to protect it from disease organisms. Evidence of spoilage may therefore point to a health hazard. For these reasons it may appear convenient to associate the two, but for a full understanding of the problem and the significance of control measures, the foodservice manager needs to observe the distinction between contamination and spoilage.

"SANITARY" VS. "CLEAN"

While we're defining terms, let's clarify what we mean by cleanliness and sanitation.

That which appears to be clean is not always sanitary. On the other hand: *That which is sanitary may not look clean at all.*

The term "clean" is a perfectly good household word, in more ways than one. We use it in this text in a general sense. And we certainly don't want to invite the wrath of dutiful parents and good housekeepers. But the technical difference between sanitary and clean facilities is not an idle one. Especially is this true for the manager of a foodservice operation, since he must have both to stay in business.

When we speak of *sanitation* in the present context we refer to conditions that are safe with respect to health. The word is from the Latin *sanitas* meaning health. Potable water, edible food, sterilized utensils, uncontaminated hands, antiseptic garbage pails, "hospital clean" floors—i.e., healthy food, healthy foodhandlers, and healthy facilities—are sanitary.

When we speak of *cleanliness* we refer generally to conditions free of soil or dirt. Healthful conditions may be implied, but for the most part we have aesthetics, good outward appearance, in mind—a face without a smudge, a glass that sparkles, a shelf wiped clear of dust, an environment free of litter or debris.

A "spick and span" surface, as we will see, may harbor harmful micro-organisms or chemicals, just as a surgically clean silver service may be splotched and water-marked. This brings us back full circle to "clean germs" and "healthy dirt." So long as we have our definitions straight, the seeming paradox will not entrap us.

A good byword for the foodservice operator is: LOOK CLEAN — BE SANITARY.

HOW WE ANALYZE THE PROBLEM

Our high school science teacher was fond of reminding us that "science is an organized body of facts, classified and arranged so as to be of practical use." In sorting out the subject of sanitation in foodservice, it is useful not only to arrange our attitudes and define our terms, but also to analyze our approach to the problem.

As suggested, we can logically look at the matter from the viewpoint of *People* —the people involved in handling food and the people whose health is in their hands; from the viewpoint of the *Food* itself—its safe condition initially and its protection in preparation and service; and from the viewpoint of *Facilities*— the sanitary condition of physical plant and equipment used in a foodservice operation.

People

The safeness of food clearly depends to a great extent on people—those who produce and process it, those who transport and distribute it, and finally the foodservice people who prepare it for the ultimate consumer. This food chain is a long one with many links, and is subject to numerous stresses on the way from farm to table. In tracing the story of food contamination, we must be aware of the farmer who over-fertilizes his field or over-dusts his crops with insecticides; of the road-weary truckdriver who makes an unscheduled stop and lets the temperature in his "reefer" rig run too high for want of electrical power; of the worker in the processing plant who adds too much chemical preservative or other adulterant; of the warehouseman who lets his guard down against rodent infestation. But we are especially concerned with the final link in the chain —the foodservice link. And, as it happens, that link is under the severest strain. Like it or not, we have to reckon with formidable evidence that it is the weakest link.

In any event we are attacking the problem from the standpoint of the foodservice manager. Let us just say he has a heavy responsibility, and a great opportunity, to help protect the public from foodborne illness. He has only limited control over the sanitary condition of food before it reaches his storeroom, but what he can do about it after it comes into the house is considerable.

In a most fundamental way, his success in dealing with the problem depends on how he handles the human factor—how he selects his people, how he trains them, and how he follows up on that training.

As we will see again and again, the foodservice manager often finds himself concentrating on elementary sanitation precepts and rudimentary rules of personal hygiene. Sometimes he will be fighting plain ignorance, as with the cook who just doesn't appreciate the danger in using a hand that has a minor, but infected, cut or burn; or a salad-maker who reaches for a knife previously used in cutting up raw chicken. At other times he will be faced simply with careless habits, as when the busboy puts down the scrub mop to add an extra place setting, without washing his hands; when the short-order waitress, working hard to keep up, serves food, handles money, clears away, makes a new set-up, catches a cigarette and serves more food—all without washing her hands.

By long-standing custom, many restaurant personnel are hired off the street, as witness the familiar "waitress wanted" sign that shows up in restaurant windows. The hard-pressed manager—indeed, the entire industry—faces a hard task in surmounting inherent defects in human behavior where good personal hygiene and sanitary habits are concerned. We can only hope that improvement in career education may serve to offset shortcomings in "upbringing" as well as in professional training.

What the manager can do about the *unsanitary customer* is a problem unto itself. The kind and frugal people who "hate to see food wasted" and put their handled but uneaten roll back into the breadbasket, and the customer with an

uncontrolled and unshielded cough are not easy to cope with. In all, it would seem that today's foodservice manager must assume the role, among other things, of a wise and discreet leader for a considerable number of his fellow men.

Food

Green plants use the radiant energy of the sun to generate organic compounds from basic earth chemicals—essentially carbon, hydrogen, oxygen, and nitrogen—and the first link in the food chain is forged. This elemental process (photosynthesis) which fashions the building blocks of life is vitally necessary to the continuation of existing plant and animal life forms evolved over billions of years. Virtually everything that we eat is organic material, composed of living or once living animal or vegetable matter. The wondrous sequence of biochemical events involved in this conversion—from fundamental elements to amino acids, to proteins, to more complex organic molecules, to human tissue—is more than our present study can accommodate. We shall have to be content with following the chain of events *after* organic compounds have become edible plants and animals in our food supply, with particular attention to what happens to food after it reaches the foodservice point. This will be adequate for our purposes, since it is well established that relatively few food disease outbreaks are traced to sources of supply.

It is hardly surprising that food ready for our consumption is a very desirable nutrition medium for other life forms, large and small. The metabolic changes which occur in the human digestive process—e.g., fermentation—are also occurring continuously outside it, and often to our detriment. This is one way of describing contamination of food that leads to foodborne illness.

As we will see, micro-organisms in food cause illness in two ways: some bacteria are disease germs in themselves which feed on the nutrients in certain foods, especially those containing milk and eggs and other foods of animal origin, and multiply very rapidly at favorable temperatures. These germs are simply using the food as a medium for growth and for transportation to the part of our body which they *infect*. Other bacteria are not infectious in themselves, but discharge toxic wastes from their own bodies which *poison* us. These toxins cause illness in much the same way that any toxic chemical does.

From ancient times man has apparently had correct instincts about how to protect himself from contaminated food. To preserve game he sun-dried it, salted it, smoked it, even chilled it when he could. To improve its flavor and texture—and with some acquired sense of the purifying effect of fire—he often cooked his food. We now know, a bit more scientifically, the value of temperature control in keeping food and rendering it safe.

Micro-organisms require a moist, warm, nutritious environment to prosper. Their growth can be slowed or stopped by refrigeration, and they can be destroyed by heat.

It is that temperature middle-ground between 45° and 140°F (about 7° - 60°C) that we have to look out for. We will do well to fix in our minds an image of this segment of the thermometer scale early in the game. It represents the danger zone in which bacteria can replicate from a single organism to billions in a matter of hours.

In our analytical approach to the problem of unsafe food, it is abundantly clear that the condition of the food itself has to be brought into central focus. What happens to it after we get our hands on it may be even more critical.

Facilities

Lastly in our analysis—but certainly not last in importance or time sequence —we view the problem from the standpoint of the safe food environment.

The design of the house itself and the sanitary features of major units of the physical plant—in a word, their "cleanability"—loom large in the architectural and engineering plans for a foodservice establishment. With forethought in these matters we have at least set the stage for a safe operation.

Chapter 7 treats this subject of "built-in" sanitation in terms of materials, design and construction, and with reference to installation and layout of equipment—all with a view to facilitating cleaning and maintenance, and eliminating entry-ways and breeding places for bacteria, insects and rodents. Adequacy of utilities and services, and other environmental factors in safeguarding people and food products, are also covered.

Subsequent chapters of Part III present detailed procedures for cleaning and sanitizing food preparation areas, equipment, utensils, and serving-ware, and provide guidance in scheduling these operations, as well as practical information on pest control and chemical hazards.

IN SUMMARY

The "mission" of a foodservice establishment becomes more sophisticated as the public grows increasingly aware of the vital importance of food to good health.

Where sanitation is concerned, the foodservice manager has an ethical and legal responsibility to protect people against illness from contamination of food by harmful organisms, their toxins, and other poisonous materials.

Technological advances in food production and marketing techniques are accompanied by potentially great risks to public health. It is not that the individual worker is any less knowledgeable or careful than he once was. He is probably more so. But in mass food processing, central kitchens, etc., a single human error can take a tremendous toll.

Bacteria are everywhere. The first imperative in the fight against foodborne illness is to awaken foodservice personnel to this supreme fact.

It is relatively easy to recite the commandments of foodservice sanitation: Clean hands — Guard food from coughs and sneezes — Serve it hot—Keep it cold. It is quite another thing to get food preparers and servers to observe these rules instinctively.

As more people eat away from home and the potential hazards mount, government agencies grow more watchful. The absolute number of foodborne illness outbreaks reported is on the rise. Whether this does or does not represent an increase relative to the size of the dining-out population, official statistics are more complete and more meaningful and have led to tighter food sanitation laws.

Sanitation regulations and public health inspections are essential, but ultimately the problem bucks down to the man at the stove, in the pantry, and on the service line. Today's enlightened foodservice manager will, on his own initiative, set high standards of sanitation and support them with vigorous training and self-inspection programs.

Clean is sanitary? Not necessarily so! It behooves the foodservice sanitarian to keep his concepts clear and his terminology straight if he is to see the problem in its true perspective and attack it with telling effect. The technical difference between food contamination and food spoilage is another case in point: Contamination is the dangerous shoal, spoilage *may* be the marker buoy.

This textbook analyzes the problem of sanitation from the standpoint of People, Food, and Facilities. This is a logical way to look at it, and one which should help you in the important task of doing something about it.

A CASE IN POINT

T. M. is the Typical Manager of a medium-size family restaurant. Table service. Counter service. Nothing fancy. Tim, we'll call him, has recently taken over. He has some firm ideas about how to "make something of the place." He is not a gourmet cook, but has had some back-of-the-house experience, a year or two out front, and knows what he wants to do: "Run a clean place, and serve good food."

Tim was brought up to respect cleanliness and good order. He has heard about food poisoning outbreaks at community picnics, and has a natural distaste for dirty kitchens. He figures that what you have to do to prevent contamination is use top-quality materials and keep the place clean. He has always cooperated with public health inspectors and followed their advice.

The restaurant has been a fairly successful business, and has only been cited for an occasional minor health infraction — never been a known source of disease. Tim doesn't anticipate any trouble. He will run a spotless kitchen, a bright, attractive dining room staffed by clean and courteous waitresses, and the germs will take care of themselves.

Is Tim's attitude toward sanitation a good one?

Not bad. His is the traditional approach of the American restaurateur. With moderate luck it has worked. Tim is all for cleanliness, but his method may not, or example, head off cross-contamination and may not avoid the dangers of holding food too long at unsafe temperatures. Surely it is preferable to know some of the *whys* of sanitation and to minimize the element of luck. And that, we submit, is why you are reading this book.

MORE ON THE SUBJECT

For further reading the student is referred to the following sources described in the Bibliography, Appendix B.

Reference 2 *Food Sanitation* (Guthrie, 1972), Chap. 1.
3 *Food Poisoning and Food Hygiene* (Hobbs, 1968), Introduction.
4 *Quantity Food Sanitation* (Longrée, 1968), Preface, Chap. 1
5 *Sanitary Techniques in Food Service* (Longrée/Blaker, 1971), Part I.
6 *Microbiology for Sanitary Engineers* (McKinney, 1962), Chap. 1.
23 *Food Service Sanitation Manual* (USPHS, 1962).
25 *Protecting Our Food* (USDA, 1966), "Challenges".
26 "Emerging Foodborne Diseases" (Bryan, 1972).

CHAPTER 2

The Micro-World

"Wine is the living blood of the grape; it is liable to sickness and doomed to death."

So wrote the encyclopedist a few years ago in a scholarly article on wine-making.[1] Little wonder that his treatise was a mixture of the scientific and the poetic, since the chemistry of wine is a complex subject and men have lyricized the grape for many centuries.

But our scholar might well have said the same about all food.

We live in a world swarming with microscopic life. Since these organisms are so small that we cannot see them, they escape our casual notice, even though we may know they are there. (Considering their importance to life and health, this may seem strange in a technically oriented society. But we can take some comfort in the fact that the existence of microbic life has not been known for very long: It was only in the late 19th Century that Pasteur convinced the French Academy of the bacterial origin of infection, and that Lister in England demonstrated the need for antiseptic surgery.)

[1] André L. Simon, *Encyclopaedia Britannica,* 14th ed., 1929

We in foodservice have every good reason to be continuously aware of micro-organisms, however, and that will be one of our primary objects at the outset. Armed with a knowledge of the micro-realm, we will be prepared to accept our responsibility, under law, to protect the consumer, and to meet our professional obligation to serve him safe, wholesome food.

We begin the present chapter with a look at some things we already know, and proceed to a closer view of the principal forms of micro-biological life which concern us in preparing, holding, and serving food. We will:

- **Be introduced to bacteria, the most common microscopic agents at work in the things that touch our lives through food, drink, and other contacts with the environment.**

- **Get an idea of the size and shape of bacteria, and how they multiply.**

- **See what kind of environment bacteria thrive in—how temperature, moisture, and nourishment affect their survival and growth.**

- Discover that bacteria and other micro-organisms convert food, spoil food, poison food, and ultimately change all food if allowed to multiply.

- Take a look at other forms of micro-life—yeasts, molds, viruses—that affect, and infect, our food.

In the beginning, God made more than the earth and Adam and little green apples. He made the heavens, of course, and the stars and planets in the deep void of outer space. He made, also, the "internal cosmos," the microscopic world of inner space.

Ever since man was put on this planet he has co-existed with multitudes of minute creatures he could not see, much less understand. We call them germs, "bugs," microbes (from the Greek *micro* meaning "small," and *bios* meaning "life"), and by some more elegant names.

In modern times we have been able to look in on these tiny life forms, but only with the aid of a microscope or other magnifying device. They are our unseen companions from the moment we begin our own existence as a small organism until we return to the dust of their micro-world.

Certain micro-organisms are native to our skin, our mouth, and our digestive tract. They are in our food and water. They are, quite literally, everywhere there is life.

We are actively involved with very few of their entire number — less than one per cent — but in the total scheme of life micro-organisms play a vital role. They are active in the biochemical changes that convert our sewage, break

down (bio-degrade) our garbage, decompose our grass cuttings and other rubbish, and turn these wastes back to the life-giving soil.

Maybe we don't give them enough credit. Can you imagine what our environment would be like if the debris of modern society did not revert to the soil? If leaves and deadwood continued to accumulate on the forest floor? Micro-organisms and larger living things are all part of nature's eternal recycling process. That of course is what ecology — the balance of life — is all about.

Micro-organisms are in every respect dynamic creatures — living, growing, working, and reproducing themselves. This is an essential point in considering their effect on foods and how we can control them.

Why not just simply avoid them, or get rid of them if they succeed in invading our systems through water, food, or dust particles in the air? This we do, insofar as we can, when it is desirable. That's a big "if", but remember that some microbes are good for us, some are bad for us, and most of them are almost completely indifferent to us.

Even among those which are not indifferent to human life, the good or bad character of a micro-organism may be purely a matter of our own option at the time. Take milk. A disease germ in milk is a villain — no question. But an organism that only sours milk may be a bad actor or a good one depending on whether we want to drink it or make cottage cheese.

It is not our purpose to pass out academy awards to the top performers, but it is instructive to note some of the beneficial roles played by microbes. In addition to making cheese, micro-organisms serve us in the fermentation process which makes bread dough rise, turns grape juice into wine, turns cereal mash into beer, and turns apple cider into vinegar.

These are, for the most part, controlled processes involving the conversion of starch and sugar into alcohol through an intermediary substance (enzyme) produced by the organism. But when "the living blood of the grape" turned to vinegar after years in the bottle, something had got out of control. For example, the cork dried out, and the air — with some bad actors aboard — got in.

As it happens, the bad guys are the performers we are mainly interested in, but before zooming in on particular types which contaminate food and cause disease, let's have a look at the general types of micro-organisms and their main characteristics.

BACTERIA

Of all forms of micro-life, bacteria are probably of greatest concern to us. These organisms occur in a variety of shapes and sizes, and are considered a form of plant life.

Size

We have said that bacteria are extremely small. To get an idea of how small, imagine 83 football fields placed end to end with a football sitting in the middle of this 8,300-yard array. If we shrink the whole scene down to a one-inch length, the football will be the size of one bacterial cell. A typical bacterium measures about 1/25,000th of an inch, or one micron (a millionth of a meter). The actual size will vary with the species of the particular cell, with the stage of its development, and with the conditions in the medium in which it lives.

Where They Are Found

Bacteria can live anywhere a man can live — and then some. They survive hotter and colder temperatures and a wider range of atmospheric conditions. Generally speaking, they thrive in a warm, moist environment that is neutral or slightly acid, but some species tolerate, even prefer, extremes of heat or cold, and some survive in a medium of high acid or salt content. Others thrive only in oxygen-free surroundings. It is safe to say that we can find some bacteria in all

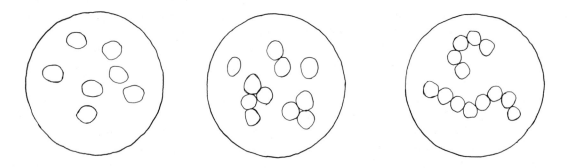

Fig. 2 Cocci singly, in clusters (staphylo-), and in chains (strepto-)

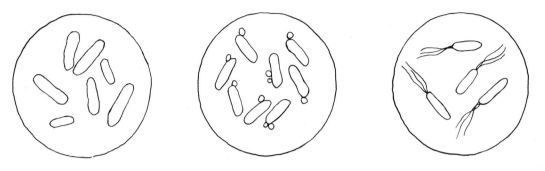

Fig. 3 Bacilli vegetative, with spores, and with flagellae

but the most highly protected or the most extreme environment. Bacteria have little or no means of locomotion, so they depend on someone or something to move them from place to place: water, food, wind, insects, rodents or other animal carriers, including man.

Shape

Since they are so small, the shape and form of bacteria may seem unimportant, but a mental picture of sorts will help us to distinguish them in a general way. Very simply, we can visualize bacteria as occurring in three basic shapes. Some are almost perfect spheres (called cocci), some are essentially rod-shaped (bacilli), and some are curled (spirilla). As they multiply, some remain separate, distinct cells; others develop in pairs (diploid); some are found in bunches like grapes (staphylococcus), and some appear in chains (streptococcus). You have almost certainly had some of the diseases caused by bacteria, and now you know where they got their names, like "strep" or "staph" infections.

Multiplication

Bacteria are one-celled, thin-walled living creatures that need food, give off wastes and reproduce. Their reproduction process is simple: They divide in two. If conditions are right, this doubling by fission may occur several times an hour. In this way, a single cell could multiply into millions of offspring in a brief time and develop into a *colony*. Colonies of bacteria may be visible to the naked eye, but do not always have a dis-

Fig. 4 Spirilla, spiral and curved

Fig. 5 A typical bacterial cell multiplies by dividing.

tinctive appearance which could make it possible to identify them. The slime we see on damp surfaces may be a manifestation of bacterial growth.

Some bacteria are able to protect themselves by forming spores. Sporulation occurs when the environment becomes too hot, too cold, or too dry, and when nutrients are scarce. The spore is a heavy-walled, weather-resistant, seedlike form of life. The bacterial spore is not a phase in the reproductive cycle, as with some micro-organisms, but a mechanism of survival. A bacterium in its growing, non-spore stage is called a *vegetative* cell.

Bacterial Growth Patterns

In the aggregate, the bacteria with which we are concerned assault a variety of foodstuffs. Some thrive on meats but may only survive on vegetables, and in other instances the reverse may be true. Certain strains of a species of bacteria may be highly selective.

In the process of living and multiplying, bacterial cells discharge waste materials, some of which are poisonous to people and even to other bacteria. These wastes are often characterized by a powerful odor. The repulsive odors associated with human feces are caused by the bacterial breakdown, in our intestines, of food materials not taken up by the body. A particle of human feces the size of a green pea contains many millions of living bacterial cells.

The temperature at which bacteria live and grow is an extremely important control factor. Considering all types, bacteria survive a wide range of temperatures and other environmental conditions. But the majority of them fare best between 60°F and 110°F (about 15°C to 43°C). Note these temperature limits: They very nearly coincide with the limits which define ideal conditions for human life.

Some bacteria live and multiply readily between 110°F and 130°F (about 43°C to 54°C). These varieties are referred to as "thermophiles," from the Greek *thermo* meaning "heat" and *phile* meaning "love." Others seem to prosper between 32°F (freezing temperature for water) and 45°F. We call this group "psychrophiles" (Greek *psychro,* "cold";

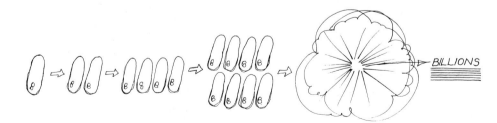

Fig. 6 Under ideal conditions micro-organisms multiply at an explosive rate, a single cell becoming billions in 10 to 12 hours.

phile, "love").

At constant temperature the survival rate for a typical colony of bacteria follows a distinct pattern.

If we touched a slice of ham with our thumb, we would probably plant several thousand bacteria on its surface. Some of these organisms probably would not survive the change in environment so for a brief period there would be fewer organisms, or no marked increase in number. This is called the *lag* phase. After an hour or so, depending on conditions, they would start to grow very rapidly. This we designate the *accelerated* growth phase.

When they have increased to such large numbers that they compete for space and nourishment, they quit multiplying so rapidly. Ultimately the colony of bacteria reaches a density at which the competition is so keen that many start dying and the total count levels off. This is the *stationary* phase— approximately as many die as are produced. By this time, our ham is so decomposed we wouldn't want to use it.

If we give some thought to everyday situations like this, we can see that they offer some very useful lessons for the foodhandler. For example:

BACTERIA NEED FOOD AND MOISTURE.

This is a key to food preservation. Many food products are handled dry, such as sugar, flour, some meats, some fruits. So long as they are kept dry, these supplies will store safely, even though some bacteria are present. But once the food becomes wet or damp, the organisms start growing, and our problems begin.

COLD RETARDS BACTERIAL GROWTH.

Knowledge of the effects of time and temperature on bacteria can be a valuable tool in keeping food safe. Bacteria grow very slowly at lower temperatures. They are not killed by freezing, but their growth stops. They grow most readily between room temperature and a little above our body temperature. Above 110°F many will start dying, and by the time we reach 130°F most will be killed. Spores are not always killed by high temperatures — they can even survive boiling — but vegetative cells are destroyed by high heat.

These facts are used in day-to-day working with foods. We should freeze foods to preserve them for long periods of time, refrigerate (chill) foods for shorter keeping periods, leave them at room temperature only very briefly, or hold them at high temperatures before serving time.

Up to this point we have talked about bacteria. Now let's consider briefly the other major classifications of micro-organisms which concern the foodservice manager.

YEASTS

Yeasts are one-celled, plant-like organisms which have a few undesirable habits but are largely beneficial to man. They thrive on foods rich in sugar, producing carbon dioxide and alcohol in the process. That is good in leavening bread and, as we have seen, in fermenting wines and beers. But to our dismay, yeasts sometimes sour food in which that flavor is not desirable.

Unlike bacterial contamination, yeast spoilage of food is easily detected — by the presence of bubbles and by the flavor and odor of alcohol. In any case, yeast damage is rarely harmful to humans.

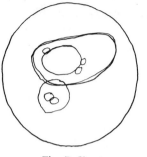

Fig. 7   Yeast

Fig. 8   Mold

When undesirable, the organisms can be killed by heating to 136°F (about 58°C) for 15 minutes.

Yeasts multiply by "budding," a kind of cell division. Under the microscope, the budding organism looks a good deal like a plant which is putting forth a new leaf. (See Fig. 7.)

MOLDS

The furry-looking stuff you see growing on bread and cheese is probably mold. So is the blue or green powdery covering on citrus fruits.

In common with some of the other micro-organisms, molds are useful (in the cheese-making process), and they are harmful (in imparting a musty odor and flavor to foods). It was once thought that molds were otherwise harmless, but now it is known that some types produce toxins which may cause illness.

One characteristic which makes molds a nuisance and a hazard is their great adaptability. They can grow on almost any food, at almost any storage temperature, under almost any conditions — moist or dry, acid or non-acid, salty or sweet. This flexibility accounts in part for the frequent appearance of mold growths in refrigerators and on damp fabrics.

Molds (Fig. 8) have a cottony look because they are made up of tiny threadlike bodies called hyphae. The hyphae produce enzymes used to digest food, and they also form spores. These spores are not survival devices, as with bacteria, but are for purposes of reproduction. The vegetative cells and spores of most molds can be killed by heating to 140°F (60°C) for 10 minutes. Some types require more rigorous treatment.

VIRUSES

Viruses (from the Latin for *poison*) are the smallest and most simply constructed of the micro-organisms. Basically, a virus is made up of a slug of genetic material in a protein wrapper. The organism is so small — about one-fifth the size of a bacterium — that it can't be seen with an optical microscope, but must be magnified electronically. Because of the processes involved in electron microscopy, that means that no one has ever seen a living virus.

In contrast to other forms of microscopic life, viruses can be *carried* by food but do not *grow* in food. They must be within a living cell to mature and multiply, and once they gain entrance they somehow order the cell to stop its ordinary life processes and, instead, produce more viruses.

We usually think of diseases caused by viruses as being passed on by poor habits. A carrier, who harbors the disease but doesn't necessarily show the symptoms, may pass the viruses from his body and contaminate his hands. Then he handles food without proper washing and infects the people who eat it. Since viruses are eliminated from the body, we can expect to find them in raw sewage and polluted waters. For example, shellfish taken from sewage-tainted waters have been found to carry the hepatitis virus. Insects and rodents which come in contact with sewage are other examples of viral disease carriers. Little is known about the ability of animal viruses to infect man, but it has been shown that some do cause human disease.

Some viruses can survive the high temperatures that destroy other forms of micro-life. They are, therefore, the most resistant to sanitation techniques in foodservice operations.

IN SUMMARY

For billions of years, and long before the arrival of man, the earth has been blanketed with microscopic life. It was only in the last century, however, that we began to understand the role of bacteria in causing disease. Some knowledge of the micro-world is necessary to the foodservice manager in understanding how to protect food from contamination and prevent foodborne illness.

Micro-organisms are everywhere — in our bodies, in the soil, in water, and in the air. They perform useful functions — as instruments in decomposing debris and in producing certain foods. But their effects are obviously harmful when they serve as agents of foodborne disease and food spoilage.

There are four main classes of micro-organisms with which the foodservice manager is concerned. The greatest menace to food comes from *bacteria*. Except in large colonies they can only be seen under the microscope. While bacteria can survive a wider range of environmental conditions than man, they flourish at temperatures which very nearly coincide with those most favorable to human life. Bacteria are slowed by low temperatures and dry conditions, and are destroyed by high temperatures.

Yeasts and *molds* are a common sight on certain foods, being easily visible in large growths. Although yeasts are helpful to man as agents in fermentation and leavening, they detract from the flavor of some foods. Molds are highly adaptable organisms which are valuable in cheese making, but some varieties are suspected of contributing to disease.

The smallest of the micro-organisms, viruses, can multiply only in living cells. For these tiny predators, foodstuffs serve largely as a means of transportation to the disease victim.

All micro-organisms are subject to destruction by sufficiently high temperatures. Viruses have the highest resistance to heat.

A CASE IN POINT

"Bacteria, yeasts, molds, viruses, pathogens, cocci, bacilli, diploid, staphylococci, streptococci, endospores, psychrophiles. What do they think I am — a microbiologist?"

Tim rebelled at first but finally settled down to the problem and attacked it with a will. A few Latin terms wouldn't stop him. In a matter of hours he dog-eared the glossary memorizing new words. He read and reread the chapter, reviewing the various types of microscopic organisms, their development stages, reproduction cycles, multiplication rates as a function of temperature, etc.

It seemed an endless task but he plunged ahead, did some additional reading (only to find it more technical), even hunted down other references to get answers the text didn't give him.

By the time he got to Chapter 3 Tim had little room in his head for *Clostridium perfringens* and *Staphylococcus aureus*. He was beginning to wonder if food sanitation was worth the battle.

Comment?

As with Tim, we're about out of breath. He is on the right track, all right, but needs to pace himself. A foodservice student must, of course, master the basic vocabulary of a technical subject like sanitation, but a microbiologist he need not be. The present text seeks only to give the typical manager enough microbiology to enable him to combat the disease agents in food — a relatively small, and manageable, number compared to the bewildering variety of micro-organisms that abound in nature.

MORE ON THE SUBJECT

For further reading the student is referred to the following sources described in the Bibliography, Appendix B.

Reference 1 *Control of Communicable Diseases in Man* (Gordon, 1965), Definitions.
2 *Food Sanitation* (Guthrie, 1972), Chaps. 2-4.
3 *Food Poisoning and Food Hygiene* (Hobbs, 1968), Part I.
4 *Quantity Food Sanitation* (Longrée, 1968), Chaps. 1, 2.
5 *Sanitary Techniques in Food Service* (Longrée/Blaker, 1971), Parts I, II.
6 *Microbiology for Sanitary Engineers* (McKinney, 1962), Chaps. 1-4.
25 *Protecting our Food* (USDA, 1966), "Processing — A Prime Protector"
26 "Emerging Foodborne Diseases" (Bryan, 1972)

CHAPTER 3

Contamination and Foodborne Illness

We all know that people sometimes get sick from what they eat. It doesn't happen very often to any one of us, and when it does we may not get very sick or be sick for very long. Moreover, in many cases we are not even sure that food was the offender, so we are inclined to shrug it off with the thought that maybe we ate too much.

But we also are aware that sometimes large numbers of people are stricken with "food poisoning" in public eating places and at community picnics — with *serious* results. These events are much more likely to be reported, as is, of course, the sensational incident in which death or grave illness is caused by contaminated food.

Statistics on the number of meals which Americans now eat away from home — almost one out of every three — tell an impressive story. This trend in national eating habits means that more food is being served under public regulation. And this in turn has led to more complete and more reliable morbidity data (information on the incidence of foodborne illness and its causes).

Statistics on foodborne illness outbreaks indicate that food contamination poses a potentially serious threat to public health. With the scientific and technical advances of recent decades, one might think that the biological agents which invade our food and menace our health should long since have been conquered. Unfortunately, this is not the case.

Around the turn of the century most restaurants were rather small and offered a limited menu. Nearly all food was prepared "to order." Provisions were often bought at the market or from a peddler early in the morning, and cooked and served within a few hours. There was little need to hold food once it was prepared, and micro-organisms did not have time to do their nefarious work.

Our life style has changed with increased urbanization and these changes have seen the advent of more and larger eating places feeding greater numbers of people. These developments have made it necessary to pre-process more food products, and to prepare and distribute food far in advance of serving time. In fact, it is not uncommon nowadays for foodservices to take delivery of supplies only once or twice a week. Along with this added flexibility in food preparation have come new opportunities for disease agents to invade and multiply in food.

Great strides have been made in food preservation methods such as freezing, drying and vacuum packing, but some of these new methods carry hidden risks that can only be avoided by careful food-handling all along the line. Improper thawing and reconstitution, for example, provide ideal conditions for bacterial growth.

One hears terms like "potentially hazardous" and "perishable" rather frequently in reference to safe food-handling and storage. The stability or perishability of foods is a matter of degree. Perishable food is defined as any food which, because of its nature or environmental conditions, is subject to biochemical change from the action of natural enzymes or enzymes secreted by invading micro-organisms.

As we have seen, the important factors are moisture content, temperature and, depending on the type of organism present, the presence or absence of oxygen in the atmosphere.

"Non-perishable" food is generally defined as food that can be stored safely for long periods of time. Most dried foods — beans, sugar, flour, noodles, and the like — fit this category. "Semi-perishable" foods, which can be stored for a few months to a year, include frozen

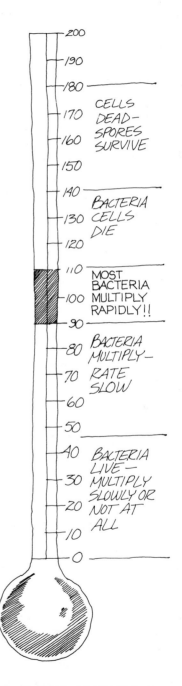

Fig. 9 **Effect of temperature on foodborne disease agents.**

products, canned goods, potatoes, and some fruits and nuts.

"Potentially hazardous" foods are perishable foods capable of supporting rapid and continuous growth of infectious or toxin-producing micro-organisms. These include foods that consist, wholly or in part, of milk or milk products, eggs, meat, poultry, fish, and shellfish.

The Seven Capital Offenders

How does food become contaminated with harmful micro-organisms and other poisonous substances? In terms of the people who handle it, the tools they use, the dangerous practices they follow, and the food and other materials they handle, the principal offenders are:

- Infected food handlers
- Poor personal hygiene
- Contaminated food supplies
- Unsafe food handling
- Unsanitary equipment
- The customer
- Hazardous chemicals

THE BIOLOGICAL HAZARD

The sources of foodborne illness can generally be classified as biological, chemical and physical. Of these, the biological hazard is by far the worst, accounting for an overwhelming majority of foodborne illness episodes. Within this category, the broad classification of micro-organisms known as bacteria constitutes the most common hazard. Pathogenic foodborne bacteria are generally considered in two sub-categories: those which cause *infection* in the host organism (e.g., our body), chiefly by their presence in very large numbers; and those which emit poisonous wastes (toxins) into the food. The latter type of disease is technically described as *intoxication*. Popularly, it is called "food poisoning." Whether by infection or intoxication, bacteria do not cause illness unless they are allowed to reproduce and increase their numbers at an explosive rate.

A relatively small group of pathogenic bacteria — numbering three or four types, depending on how morbidity reports are interpreted — account for more than 90 per cent of the diseases in man associated with food. We will discuss in some detail three principal offenders and make brief mention of others. This should suffice for our purposes, since the control measures necessary for managing these leading causative agents will serve to safeguard the consumer from the remaining, lower-incidence pathogens.

Staphylococci

Highest on the list of offenders in the intoxication category of disease organisms is the *Staphylococcus aureus* species of bacteria. Staph organisms thrive in the presence of atmospheric oxygen. They are commonly found in the nasal passage and throat of humans — even in people who are in good health. They are also normally present on the hands and skin, especially in infected cuts, abrasions, burns, boils and pimples. It can readily be seen that staphylococci are a "people-associated" problem.

Food products frequently involved in

staph intoxication include cooked ham and other meat products, stews, gravies, pastry fillings, and other moist food components.

Consider how a staph problem could arise in your kitchen. Assume that the chef has a skin infection on his neck. When it itches, he naturally scratches himself. Let's say one of his assignments is to prepare chicken for use in chicken salad, an item usually held for some time before being served. Once the chicken is ready, it will not be cooked further.

Fresh from the pot, the chicken is carved by the chef, who plants staph germs from his neck every time he touches the bird. While waiting to be used, the sliced chicken should have been quickly chilled, which would have stopped or slowed bacterial growth. Instead, it is put aside on a work table to endure bacteria-incubating room temperatures. Next stop is the salad pantry where the chicken is further handled — chopped or diced — and mixed into the salad base, itself prepared separately and held at room temperature. All the while — from slicing to mixing — the population of staph germs is growing at an astounding rate. The completed mixture, staphylococci and all, is once more set aside, unrefrigerated, to await the customer's call for a chicken salad sandwich.

Even if the chef had not been infected, staphlycoccal organisms could have entered the salad at any stage in its preparation. The knife used to dice the chicken could have been contaminated through prior use on raw foods in which staphlyococci are naturally present. A waitress entering the kitchen could have coughed or sneezed over the food. The chicken could have been sliced on a chopping block so worn from use that it had plenty of crevices in which bacteria could lurk. Some of the ingredients of the salad base might not be ideal media for bacterial growth, but they could have carried the organisms to the chicken, where conditions were highly favorable.

The rest of the story is sad and unpleasant to relate. Some two to six hours after ingesting this poisoned food, the customer suffers a sudden onset of nausea, salivation, vomiting, diarrhea, abdominal cramps, sweating and dehydration, weakness and prostration. This description is not merely a dramatic representation of what might have happened. These are symptoms in a typical case documented by the U. S. Center for Disease Control. Morbidity rates indicate that the customer can be expected to recover, but only after agonizing discomfort lasting from one to two days.

In combatting the danger of staphylococcal poisoning, particular attention should be given to the handling of potato, poultry, and fish salads. Salad dressings, high-protein leftovers, and sauces low in acid must also be watched.

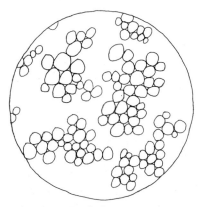

Fig. 10 *Staphylococcus aureus*

Clostridium Perfringens

The second-most common offender among disease-generating bacteria is *Clostridium perfringens,* a spore-forming organism which thrives in the *absence* of oxygen. It is found in soil, dust, and the intestinal tract of animals. These pathogens are therefore likely to accompany almost any food product brought into the house. Since they form spores, they survive moderate cooking temperatures, including boiling water (212°F, 100°C).

C. perfringens is associated with cooked meat and poultry that has remained at room temperature for several

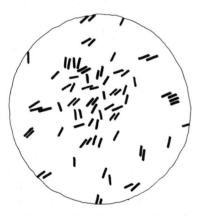

Fig. 11 *Clostridium perfringens*

hours, or has been permitted to pass slowly through these incubating temperatures as a result of gradual heating or cooling. Too slow heating or cooling is a particular problem in the interiors of large food masses. The disease caused by *C. perfringens* is called "perfringens food poisoning."

Consider a typical case. Let's say we are preparing a large pot of vegetable soup, and that it will contain fresh car-

rots. The soup is to be boiled in a 20-gallon stock pot on Wednesday evening and served for lunch on Thursday. After cooking, it is placed on a side table, at room temperature, until cool enough to be refrigerated. The soup is removed from the refrigerator at 10:30 Thursday morning and brought to the steam table on the serving line.

Look what we have done. Carrots are grown in the soil, of course, where they very probably have been contaminated with perfringens spores. It is impossible, using normal methods, to remove all these spores by washing and scraping. The soup was boiled but this does not kill the spores. A large quantity of material has been held in the danger zone for many hours. This allows the bacteria and their spores plenty of time to vegetate, grow, multiply, and poison the soup. When the soup was brought to serving temperature on the steam table it was not heated enough, or was not brought through the danger zone quickly enough, to destroy the bacteria. In fact, the organisms were given another opportunity to germinate and multiply.

Clostridium botulinum, a close relative of *C. perfringens,* is probably the best known of all food-poisoning organisms because of its extremely lethal characteristics and the sensational publicity accorded to outbreaks of botulism. This disease is truly a scourge when it strikes. Statistics indicate that 65 per cent of botulism cases are fatal. But the fact is that botulism outbreaks originating in commercial foodservice establishments are exceedingly rare. *C. botulinum* is largely associated with improperly processed, usually home-canned, low-acid foods (e.g., green beans), smoked vacuum-packed fish, and certain food products of fermentation. It is only fair to say, however, that in the case of the contaminated vegetable soup described above, there was a chance that the incriminated carrots carried botulinum as well as perfringens to the soup kettle. Although statistically remote, the possibility of botulism is not one for the foodservice manager to dismiss lightly.

Salmonella

The third disease organism of major concern to us, Salmonella, is in the other

Fig. 12 A species of salmonella

broad category of pathogens, which cause illness by multiplying profusely and *infecting* our digestive tract. There are more than 1,300 specific types of salmonella, and 50 of them occur commonly. Salmonellae do not form spores, and are supported by an environment in which oxygen is available. Typhoid fever is caused by one type of salmonella.

The primary source of salmonella is the feces of infected domestic and wild animals, and of humans. The foods usually involved are meat, poultry, eggs and their products.

It is nearly impossible to dress animal carcasses without getting fecal matter on the meat. Salmonellae are readily killed by normal cooking methods so long as all portions of the product are heated sufficiently to 145°F (about 63°C).

In dealing with the salmonella problem, the foodservice manager must be mainly concerned with people, and with what is referred to as *cross-contamination*. For example, if a cook is deboning a cooked turkey and uses the same cutting board or knife that was used to trim the turkey before cooking — without properly cleaning and sanitizing the board or the knife — the trouble can be anticipated. It is highly probable that the raw turkey was contaminated with salmonella.

The carrier — a person with disease-causing organisms in or on his body, who isn't himself sick — was important when the staphylococcus organism was discussed. He is equally important, if not more so, in the case of salmonella.

Shigella

Another significant foodborne illness of bacterial origin is shigellosis, commonly called "bacillary dysentery." Shigellosis outbreaks usually affect large numbers of people. Like salmonellae, dysentery bacteria cause an *infection* in their victims.

Man himself is the prime reservoir for the dysentery bacillus. The spread of this organism is attributed almost entirely to poor personal habits. Roaches, flies and rodents are also responsible for the transmission of *shigella*. Many people who have had dysentery become carriers of this pathogen for periods lasting from a few weeks to two years or more.

Table 3-1

Foodborne Illness

DISEASE	*	CAUSE
Staphylococcal 26% Intoxication Symptoms: vomiting, diarrhea, cramps. Onset in 3-8 hours. Duration 1-2 days.		*Staphylococcus aureus* — bacteria commonly present in nose, throat and skin infections; release toxins highly resistant to heat.
Perfringens 16% Food Poisoning Symptoms: nausea, diarrhea, acute inflammation of stomach and intestines in 8-20 hours. Duration up to 24 hours.		*Clostridium perfringens* — spore-forming bacteria found generally, in soil, dust, intestinal tract of animals; spores withstand most cooking temperatures; surviving cells flourish in absence of air. Natural contaminant of meat.
The General Case of gastro-intestinal illness		**Traced mainly to foodhandlers, and to foods of animal origin contaminated by bacteria which are allowed to multiply through lack of refrigeration, undercooking and insufficient reheating.**
Salmonellosis 13% Infection Symptoms: headache followed by vomiting, diarrhea, abdominal cramps, fever. Severe infections cause high fever and may be fatal. Onset in 12-36 hours. Duration 2-7 days.		Salmonella bacteria widespread in nature; live and grow in intestines of humans and animals. Some 800 types cause gastro-intestinal illness; one species, *Salmonella typhosa*, causes typhoid fever.
Shigellosis 2% (Bacillary dysentery) Infection Mild to severe symptoms: diarrhea, fever, cramps, chills, lassitude, dehydration. Onset in 1-7 days. Duration indefinite depending on treatment..		*Shigella sonnei* and other species — bacteria found in feces of infected humans, transmitted person to person and by contaminated food and water.

*NOTE: Percentages are approximate figures representing relative incidence of reported foodborne illness outbreaks based on morbidity statistics averaged over the five-year period 1967-1971.

Causes and Control

FOODS IMPLICATED	PREVENTIVE ACTION
Moist, much-handled, high-protein food left at bacteria-incubating temperatures: milk-egg custards, meat pie, turkey stuffing; chicken, tuna, potato salads; gravies, sauces. Warmed-over food.	Store these foods at or below 40°F. Exclude foodhandlers with respiratory illness, pimples, infected cuts, burns. Avoid hand contact with food. Reheat leftovers through and through to 165°F or above.
Raw meat, raw vegetables; partly-cooked meat slowly cooled and served later, or served after moderate reheating.	Careful time-temperature control: quick-chill cooked meat dishes to be eaten later. Isolate raw components that may cross-contaminate cooked menu items.
Poultry and its dressing, meats, gravy; eggs, dairy products, potato, macaroni salads, and such; custards, cream-filled pastry; fish, shellfish. (Typical "potentially hazardous" foods cited by public health authorities.)	**Time and temperature control: cook food properly and serve, hold above 145°F, or chill promptly to 45°F or below. Good personal hygiene; healthy foodhandlers. A clean kitchen; sanitary equipment and utensils.**
Meat and poultry, especially fine-cut components; egg products, puddings, shellfish, soups, gravies, sauces. Warmed-over food.	Strict personal hygiene. Avoid fecal contamination from unclean foodhandlers and unsafe operating practices. Eliminate rodents and flies.
Contaminated milk, beans; potato, tuna, shrimp, turkey and macaroni salads; apple cider. Moist, mixed foods.	Personal hygiene. Safe food-handling. Sanitary food and water sources. Insect and vermin control. Sanitary sewage disposal.

Table 3-1 (cont'd)

DISEASE	CAUSE
Botulism 2% Intoxication Symptoms: vomiting, abdominal pain, headache, double vision, progressive respiratory paralysis. Onset 2 hours to 6 days. Paralysis may persist for months. High fatality rate, 50-65% in U. S.	*Clostridium botulinum* — bacteria that grow in absence of air, as in sealed containers, and form spores with high resistance to heat. Toxins are deadly but vulnerable to high temperature. Found in soil, water and animal intestines.
Streptococcal Infections 2% Gastro-intestinal symptoms: nausea, vomiting, colic and diarrhea. Scarlet fever, septic sore throat symptoms: tonsillitis, high fever, headache, vomiting, rash.	*Streptococcus faecalis* and associated species — bacteria found in soil and manure, transmitted by meat animals and workers contaminated with feces. *Streptococcus pyogenes* — bacteria mainly transmitted through the air from nose and throat of infected humans.
Infectious Hepatitis 2% Viral infection Symptoms: jaundice, fever, nausea, abdominal discomfort. Onset 10-50 days. Duration few weeks to several months.	*Hepatitis virus A* — Originates in feces, urine and blood of infected humans and human carriers. Mainly transmitted person to person, also waterborne.
Trichinosis 2% Parasitic disease Symptoms: vomiting, diarrhea, fever, sweating, muscular pain, chills, skin lesions, prostration. Incubation 4-28 days.	*Trichinella spiralis* — Delicate roundworm, larva of which invade the intestine, later imbed themselves in muscle tissue. Transmitted by infected swine, rats and certain wild animals.
Other Illnesses 5% of biological origin	(This category includes *Escherichia coli* infection and other bacterial and para-
Chemical Poisoning 6%	(Including illness caused by toxic chemicals in pesticides, cleaning compounds, solvents, polishes, and other non-food
Diseases of Unknown Cause 24%	(Etiology undetermined because food

FOODS IMPLICATED	PREVENTIVE ACTION
Improperly canned or refrigerated low-acid foods: green beans, corn, beets, spinach, figs, olives, tuna. Smoked fish, fermented foods.	Pressure-cook foods at high temperature in canning. Boil and stir home-canned foods for 20 minutes before serving. Keep foods refrigerated. In salt-curing food, use plenty of salt. Discard food in swollen cans.
Sausage, evaporated milk, meat croquettes, meat pie, poultry, ham, pudding.	Chill foods rapidly in small quantities. Cook food thoroughly. Prevent fecal contamination by foodhandlers.
Milk, ice cream, eggs, steamed lobster, potato salad, egg salad, custard, pudding.	Pasteurized dairy products. Rapid chilling, thorough cooking. Exclude foodhandlers with known strep infections.
Shellfish, harvested from polluted waters. Milk, orange juice, potato salad, cold cuts, frozen strawberries, glazed dough-nuts, whipped cream.	Cook oysters, clams, etc., thoroughly. Disinfect, heat-treat suspected water and milk. Personal hygiene. Care in sterilizing syringes and needles used in injections.
Raw or undercooked pork from hogs fed contaminated swill. (Infected bear meat also indicted.)	Cook pork through and through to 150°F or above.
sitic diseases of comparatively low incidence.)	
substances; by metal poisoning from lead, copper, cadmium and zinc containers and equipment; by misuse of food	preservatives, colorants and other food additives; and by natural poisons: toxic molds, fungi, etc.)
samples not available or laboratory tests	inconclusive.)

The law in most states makes it obligatory to report known cases of shigellosis to the local health authority.

Streptococci

Organisms of the streptococcus groups, familiar in "strep" throat, scarlet fever, etc., may also infect us through the medium of our food. Their entry into food is often gained through nasal discharges from persons who have actual disease symptoms or who are carriers of the organisms. Several varieties of streptococcus found in the intestinal tract are suspected of causing foodborne illness. They are passed on to food through excreta on unclean hands.

An obvious preventive measure is to exclude foodhandlers with known streptococcic infections. Thorough cooking will destroy most varieties of these organisms and refrigerated storage will inhibit their growth and reproduction.

Other Pathogenic Organisms

There are two more foodborne diseases which bear some discussion at this point. One is trichinosis — caused by a parasite — and the other is infectious hepatitis — caused by a virus.

Trichinosis involves delicate, coiled roundworms that are the larvae of *Trichinella spiralis*. The female worm invades the lining of the small intestine and lays her eggs there. After the eggs hatch, the tiny larvae are carried by the blood to muscle tissue where they imbed themselves. The most common source of this parasite in the United States is pork. With improvements in methods of raising pork for food in the past 20 years, the number of cases has dropped con-

siderably, but the situation still bears watching.

Sufficient heat will kill the organism and its larvae. To destroy the imbedded worms, pork should be cooked thoroughly to 170°F (about 75°C) or above. Freezing will also kill the worms if the pork is stored at 5°F (−15°C) for 30 days, − 10°F (about −23°C) for 20 days, or −30°F (about −34°C) for 12 days. Trichinae also infest bear meat and certain game animals.

Not a great deal is known about the organism that causes infectious hepatitis, but it can be traced. It is a virus found in feces and urine. Food sources of primary concern are shellfish harvested from polluted waters. Also involved are milk, orange juice, potato salad, cold cuts, strawberries, and glazed doughnuts. The key items in controlling this virus are good personal hygiene, sanitary disposal of sewage, and thorough cooking.

Poisonous Plants

There are hundreds of poisonous plants in the world, most of them not related to foods, but one merits a brief mention. We refer to the fungus known as the mushroom. Poisonous and non-poisonous mushrooms often look so much alike that the untrained eye can't tell the difference. To avoid the problem, the best rule is: *Be sure of the source*. There is no sure method that the amateur can use in detecting the poisonous varieties.

THE CHEMICAL HAZARD

Chemical contamination is a matter of concern all along the food supply chain. The danger has been increased by the industrialization of agriculture and the trend toward mass manufacture and pre-processing of food products.

Pesticides

Of the chemical agents causing foodborne illness, those receiving the most attention have been the pesticides. These agents can enter the food supply by several routes:

— Pesticides are applied directly to the growing plant or animal to protect it from insect, fungus, and microbe attack. If the protective agent is not washed away or otherwise removed, it can be passed on to the consumer.

— Food animals and plants may take up pesticidal agents during the growing process and incorporate them in living cells, e.g., DDT in grain and mercury in fish.

— Food can be contaminated by chemical agents during processing and at the foodservice point.

The foodservice manager has to exercise extreme caution in the use of pesticides and germicides in any food preparation area. Hazardous chemicals may enter foods through spillage and as the result of defective containers and unsafe storage practices.

Preservatives

There have been a number of reports of illness caused by preservatives and other food additives that were used in excessive amounts. Some coloring agents are toxic in high concentrations. A phenomenon known as the "Chinese restaurant syndrome" results from too generous use of monosodium glutamate (MSG) and causes, usually, mild irritation in the consumer. Proper use of certified "food grade" materials is the obvious course for the foodservice operator to follow to avoid these hazards.

Poisonous Metals

Some metals, like iron, are necessary components of the human diet in at least trace quantities. As is true with almost any substance, however, they become toxic at excessive levels of concentration. Certain metals — notably copper, cadmium and lead — can be the source of poisonous reactions in the presence of some food materials. Food substances with a low pH value (high in acid content), such as fruit ades, carbonated beverages, sauerkraut, tomatoes, etc., react with some metals to form toxic products. High-acid foods have caused gastric upset after being stored in cadmium-plated and galvanized (zinc-covered) containers. Copper water lines accidentally exposed to carbonated beverages also have figured in poisoning incidents. For safety's sake, always use kettles and metal utensils only for the purposes for which they are designed.

THE PHYSICAL HAZARD

Physical contaminants such as chips of glass from broken light fixtures, and metal fragments from kitchenware and tableware are obvious dangers, and the foodservice manager must be on the alert to minimize these hazards. To the alert manager, a worn can opener blade is not just an annoyance. He knows that it may, as often happens, shower metal curls on the food in the can being opened. He also knows that the common — but inexcusable — practice of scooping up ice with a glass is definitely hazardous, since any chips may find their way into beverage servings. Eliminating these contaminants involves such factors as plant design and seating arrangement, as well as training of personnel in safe operating procedures.

Radiological hazards are not a major problem but are worthy of some mention. By and large, radiation intensities are kept well below any possible danger level in food preservation processes using this technique. Regulatory agencies are, however, under pressure from environmental scientists and others to keep prescribed safety standards under review.

IN SUMMARY

Official guideposts marking the way for a foodservice manager who seeks to avoid the causes of foodborne illness are many and varied. Here is a list which takes the positive approach. It reviews the 10 most important rules of safe food-handling, infractions of which were proven causes of illness over a 10-year period (1960-1970).[1]

THE TEN COMMANDMENTS OF SAFE FOODSERVICE

1. Refrigerate food properly.
2. Cook food or heat-process it thoroughly.
3. Relieve infected employees of food-handling duties.
4. Require strict personal hygiene.
5. Use extreme care in storing and handling food prepared in advance.
6. Give special attention to preparation of raw ingredients (liable to con-

tamination) which will be added to food that gets little or no further cooking.
7. Keep food above or below bacteria-incubation temperatures.
8. Heat leftovers quickly to a temperature lethal to bacteria, or cool quickly for storage.
9. Avoid carrying contamination from raw to cooked and ready-to-serve foods via hands, equipment and utensils.
10. Disinfect storage areas without contaminating the stored food. Clean and sanitize food preparation and serving equipment.

Close attention by the foodservice operator to these rules of good practice should prevent most of the diseases attributed to bacterial poisoning and infection, and a large part of all foodborne disease. The evidence is overwhelming that most illness problems originate at the foodservice point.

---

[1] Based on reports of the U. S. Center for Disease Control, Atlanta, Ga.

A CASE IN POINT

After a few days Tim rallied from his first encounter with microbes. Then he waded into the thicket of foodborne illness.

As he read along he developed a mixed feeling about the subject: a sense of awe at the complexity of the world of micro-organisms, and a sober feeling about their importance to people's health. There really is something to this business of safe food. Germs are not to be dismissed so lightly after all.

Germs, he now realized, can get into almost any kind of food, at any stage of its handling, and from nearly every imaginable source — the handler's mouth, skin, hair; the soil, air, water; insects and rodents. Even sanitary food may ultimately become unsafe. It only takes *one* germ and *time*. And virtually every kind of food carries *some* germs. *All* raw poultry is naturally contaminated.

He pondered these many hazards as he reviewed the list of diseases on the foodborne illness chart, noting the specific germ that causes each disease, the kind of food it contaminates, and how the foodservice operator can prevent contamination. He made fair progress for a while, but eventually found himself snared in a maze of detail. Some organisms poison us, some infect us. Some develop spores, some do not. Cooking kills vegetative cells but not all spores. Only extreme heat neutralizes toxins. Salt is normally a preservative but some germs like it. Oxygen stifles some cells, others thrive on it. Cold retards most bacteria; a few are cold-loving. Highly acid foods are usually safe, but not always.

Tim was on his third trip through the chart when some boldface type glared out at him from the middle of the page: "The General Case of Gastrointestinal Illness. . . ." Food germs that cause illness have similar origins; the kinds of food they flourish in have a lot in common; the way you keep them from multiplying is pretty much the same. And when you destroy the leading offenders you have very nearly destroyed them all.

Tim was encouraged. It did not even occur to him to think back to his original idea: "Just keep it clean."

What do you think — was Tim's effort worth it?   ⟶

No doubt. Tim has come full circle, to be sure. But now he knows something of the "why" of sanitation. We need not be scientists to "manage the microbe," but an enlightened approach will motivate us in the first place, and show us how to do a better job.

MORE ON THE SUBJECT

For further reading, the student is referred to the following sources described in the Bibliography, Appendix B.

Reference 1 *Control of Communicable Diseases in Man* (APHA, 1965).
2 *Food Sanitation* (Guthrie, 1972), Chaps. 3-5.
3 *Food Poisoning and Food Hygiene* (Hobbs, 1968), Part I.
4 *Quantity Food Sanitation* (Longrée, 1968), Chaps. III-VII.
5 *Sanitary Techniques in Food Service* (Longrée/Blaker, 1971), Part I.
6 *Microbiology for Sanitary Engineers* (McKinney, 1962), Chap. 3.
12 *48 Ways to Foil Food Infections* (Bete, 1970).
17 "Food-Borne Illnesses" (NRA Reference Chart).
18 "What You a Food Service Operator Should Know About Salmonellae" (Foster, 1967).
19 "Discover the Unseen World, Prevent Food Poisoning" (Mich. State Bulletin, 1966).
25 *Protecting Our Food* (USDA, 1966).
26 "Emerging Foodborne Diseases" (Bryan, 1972).
27 *Foodborne Diseases of Contemporary Importance* (Bryan).
29 *Foodborne Outbreaks, Annual Summary 1971* (USPHS).
30 "Salmonellosis" (Zottola, 1967).
31 "Staphylococcus Food Poisoning" (Zottola, 1968).
32 "Clostridium Perfringens Food Poisoning" (Zottola, 1971).

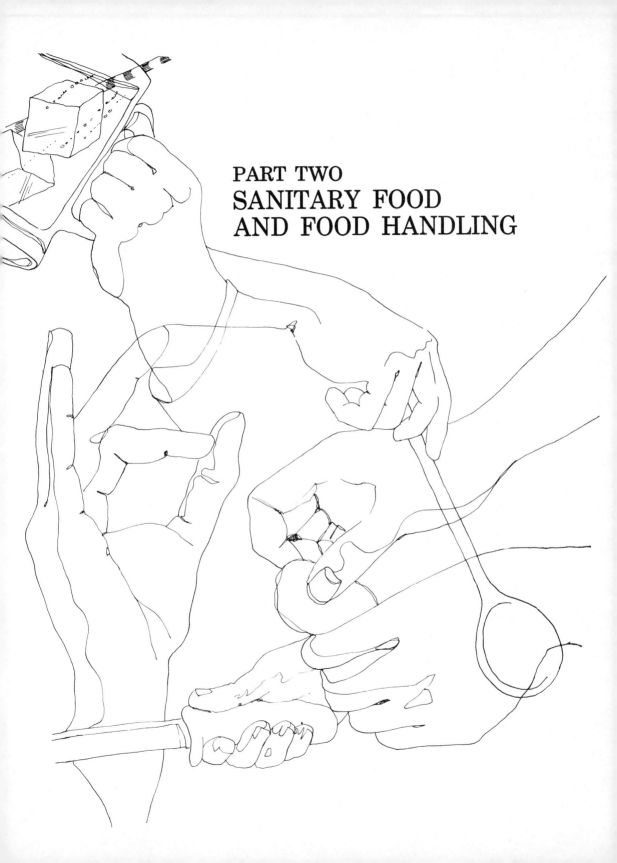

PART TWO
SANITARY FOOD
AND FOOD HANDLING

CHAPTER 4

The Safe Foodhandler

Teeming with germs both harmless and virulent, man is the single most common source of food contamination. However he touches his environment—with his hands, his breath, his perspiration—he spreads bacteria and other micro-organisms. Every unguarded cough or sneeze transmits a wave of invisible life forms capable of causing disease. Human as well as animal excrement is a significant factor in the distribution of pathogens that invade the food supply.

It is ironical that people are the culprits as well as the victims in food poisoning incidents. We do it to ourselves. The foodservice manager intent on providing safe and wholesome food is confronted with a seeming paradox: somehow he must erect a sanitary barrier between his product and the people who use it—employees, customers, even himself. To do this requires, first of all, a trained staff of safe foodhandlers.

A safe foodhandler—

- **Is in good health**
- **Has clean personal habits**
- **Handles food safely**

THE ENEMY IN AND AROUND US

An apparently healthy individual may harbor sizeable colonies of staphylococci on his hair and skin, in his mouth, throat and nose. His lower intestinal tract is not an uncommon habitat for salmonella and perfringens organisms. According to one estimate, up to 50% of all foodhandlers are bearers of disease agents transmissible by food. Another study indicates that more than half of the general public carries highly toxic strains of staphylococci.

Once the individual becomes visibly ill, his bacterial count increases dramatically—as does his potential as a food contaminator. Staph germs abound in and around boils, pimples, carbuncles, inflamed cuts, and infected eyes and ears. A sore throat, a nagging cough, the sinus pains and other symptoms of the "common cold"—these are further signs that micro-organisms are taking over, with possibly dangerous consequences for the foodservice operation. The same may be said of gastro-intestinal ailments—a touch of diarrhea, an upset stomach.

Even when the illness passes, some of

the organisms that caused it may remain, a source of re-contamination. Salmonellae may persist for months after the patient has recovered, and the hepatitis virus has been found in the intestinal tract up to five years after the disappearance of disease symptoms.

Sneezing or coughing are obvious ways in which infection can be spread, but the food worker's hands represent the most common vehicle whereby contaminants are conveyed to food from the human body and other reservoirs of disease. The transfer may involve simple personal acts that, in another setting, would at worst be considered uncouth behavior: picking the nose, rubbing the ear, scratching the scalp, fingering a pimple, habitually running hands through the hair, coughing and sneezing without benefit of a handkerchief, etc. It may, however, also involve foodhandling practices which are much more serious than rude manners. Whatever the act, it could easily be overcome by the simple act of washing the hands.

The litany of unsafe practices in foodservice is a long one, but here is a list of oft-repeated events which should always be followed by thorough hand-washing:

—Intimate contact with infected or otherwise unsanitary areas of the body.
—Use of a handkerchief.
—Hand contact with unclean equipment and work surfaces, soiled clothing, wash rags, etc.
—Handling raw food, particularly meat and poultry.
—Handling money, smoking a cigarette.
—Clearing away used dishes and utensils; scullery operations.

Fig. 13 The customer's health is in her hands.

Hands. Hands. Hands. Every one of these everyday actions, and numerous others, involve use of the food worker's hands. Frequent hand-washing is the only answer.

BUILDING THE SANITARY BARRIER

The sanitary barrier which the manager must establish is simple to describe but not always easy to maintain. It is built on (1) high standards of personal cleanliness among workers, and (2) constant supervision to ensure that only healthy workers come in contact with food, and that they do their work in a safe manner.

Emphasis on personal hygiene is a fundamental responsibility of the food-service manager in training and directing his staff. He shouldn't have to teach an employee to wash his hands after a visit to the toilet. Or to keep a daily date with the shower. These are lessons we are supposed to have learned at our mother's knee.

To pursue the subject of personal cleanliness with mature people may be an awkward and embarrassing task for the manager, but the alternative is, at the least, an unfavorable impact on patrons, which is bad for business; or, at the worst, a breakthrough for disease germs, which is bad both for patrons and their patronage.

Here are some guidelines for organizing and managing a program of good hygiene and safe handling practices:

—In selecting, placing and training personnel, give first priority to customer protection.

—Evaluate the health of a prospective employee along with his other qualifications.

—Begin personal hygiene training when the worker is hired and continue it throughout his employment.

—Provide the worker with a well-equipped lavatory, dressing room and lounge so he can observe the rules of personal cleanliness.

—Provide the right tools—equipment, work space, utensils—for safe handling of food. Make it simpler for the worker to operate without touching food.

—Provide hand-washing facilities con-venient to kitchen, pantry and counter work stations. Make it easier to work with clean hands.

—Protect customers and staff against contamination which may originate with other customers by providing for canopies, sneeze guards and other food shields in self-service arrangements, and by seeing to it that the necessary serving utensils, tongs, etc., are available to the user and are in good order.

THE MANPOWER DILEMMA

The foodservice manager is obligated to protect his customers and his employees from workers who have health problems that can affect the sanitary quality of food. Professional ethics, good business sense, and the law all say so. In most communities, the health code prohibits persons who have a communicable disease, and persons who are carriers of such disease, from work activity which may result in contamination of food or food-contact surfaces. To complicate matters, today's manager is confronted with the chronic, two-horned dilemma of untrained workers and a high turnover rate.

Obviously, the best time to find out if a worker is a poor sanitation risk is *before* he starts on the job—at the first hiring interview. But how is the food-service manager, a non-medical person, to judge the health of an applicant? He can't diagnose, of course, but he does have three aids to guide him in making a hiring decision.

First, there's the application form. In addition to the routine questions on age, sex, Social Security number, work experience, etc., it should request information on past illnesses, injuries and current health. The answers may not be wholly reliable, since the applicant will be tempted to omit ailments which might disqualify him, but it is one indicator.

Next, the manager can learn a lot simply by observing the applicant. Unobtrusively, he can apply a mental check list:

Is the worker neatly groomed and dressed? If not, it is reasonable to expect that his work habits may be questionable. Dirty clothing, unkempt hair, ragged and unclean fingernails, and an unpleasant body odor should all flash danger signals to the alert manager.

Does he show evidence of skin infections, pimples, acne, open sores? If so, he presents an obvious sanitary hazard.

Is he given to repeated coughing and sneezing? (Evidence of possible throat, sinus or respiratory infection.)

Does he compulsively pick at his face, scalp or neck? This may be a nervous habit or a sign of infection—in either case an undesirable condition for a foodhandler.

Of course the manager can and should seek medical opinion in doubtful cases. Many organizations routinely schedule a medical examination for new employees as a matter of policy, and some communities require health certificates for foodservice personnel. Mandatory or not, periodic checkups are a prudent procedure which often reveal the presence of disease organisms in persons who are only carriers and show no outward symptoms. Some county and municipal regulatory agencies administer these tests for a nominal fee or at no cost.

The problem of excluding infectious personnel does not end with the hiring process, however careful and efficient. In fact, the worker who becomes a carrier of contamination *after* he is on the job is far more likely to be the source of food poisoning incidents. Frequently the contamination originates with an established employee whose illness or infection comes and goes in a week.

The truly safe foodhandler is a product of continuing vigilance on the part of management. Every workday the conscientious manager must be on the lookout for disease symptoms in his workers and for unsafe personal habits—the telltales of events which can defeat the most ambitious program of foodservice hygiene.

SETTING UP THE RULES

All the supervision in the world won't ensure safe foodhandling unless employees clearly understand what is expected of them. Management must establish definite "house rules" that are easily understood and uniformly enforced. Workers should learn the rules from the manager in person, or from a highly dependable subordinate. The policies should be written in simple language, and posted for all employees to see—in restrooms, over washstands and lavatories, and on bulletin boards. In effect, we're waging war against an enemy that can't be entirely defeated, and the troops have to understand the battle plan. The rules and their wording may vary, depending on the size and complexity of the foodservice operation, but they should definitely cover these areas:

Clean Outer Garments

Soiled work clothes that have gone for days without laundering are bad on two counts: They're unappetizing to patrons, who may rightfully wonder if the food being served is as unsanitary as the person serving it. More significantly, dirty clothing is a repository of disease organisms. Only one hitching up of the trousers or smoothing of a dress is enough to start the contamination cycle: from the clothing, to the hands, to the food. The solution: Insist on clean uniforms, and encourage changes as often as is necessary.

Management may choose to furnish uniforms along with laundering and maintenance service, or workers may be required to provide their own. In any case, work clothes should be donned in the establishment and not worn while commuting to and from work. And street clothes shouldn't be worn in food preparation or serving areas. No surgeon would wear his gown on the way to the hospital, or his street clothes while in the operating room—and for essentially the same reasons.

Some items of jewelry and other personal decorations—a highly ornate ring or wristwatch, for example—are soil collectors and are difficult to clean. They are potential hazards and should be prohibited in foodservice areas. As a favorite breeding place for micro-organisms, the hair should be put under effective restraints, such as hats, caps, and hairnets (preferable to snoods).

Cleanliness of Workers

Dirty people are dangerous, for the obvious reason that they compound the natural contamination found even in clean and healthy people. They also are a poor advertisement for the house. In the eyes of a customer, one worker with a slovenly and unclean appearance can indict the entire operation. Touchy as the subject may seem, the manager is bound to insist that his workers bathe daily—even more often if their job requires it.

Open cuts or abrasions are equally bad for worker, patron, and public image. Wounds and open sores should be antiseptically bandaged and covered with a waterproof protector.

The most important aspect of personal cleanliness is frequent and thorough hand-washing. Body odor may offend, and lack of bathing may accelerate bacterial growth, but most often it will be dirty hands that transmit contaminants to the food product. Washing should follow any act which offers even a remote possibility that the hands have picked up contamination. And it should precede any handling of a food product, food-contact surface, or utensil. One of the most notorious incidents of food poisoning on record involved a single worker who scratched a facial infection and then handled a large amount of sliced meats.

It may seem elementary, but clean hands are so critically important that employees should be instructed in proper hand-washing procedure, if there is any doubt about their knowing how. The soaped hands should be rubbed together with a rotary motion for half a minute with special attention to the areas between the fingers and around the fingernails. A brush should be used to cleanse beneath the fingernails.

With this house policy, as with others, it is not enough merely to set up rigid rules. The reason behind a rule should be clearly explained. Rather than warning employees to keep their fingernails trimmed, or else!—tell them: "Fingernails that are long or ragged are very difficult to keep clean, and they can harbor a tremendous number of germs."

Dangerous Habits and Actions

The number of actions which can be hazardous in foodservice is virtually endless. All too often they are ingrained habits unconsciously practiced by workers, and are therefore hard to break. We have mentioned several in connection with mandatory hand-washing. Here are some others not directly associated with hand-washing requirements which will indicate the scope of the manager's problem:

—Using wiping cloths to remove perspiration. Spitting on the floor or into the warewashing sink.

—Coughing or sneezing in a food preparation area.

—Washing hands in sinks used to prepare foods.

—Smoking near food in preparation or serving areas.

—Picking up bread, rolls, butter pats, and ice with the bare hands.

—Handling place settings or food after wiping tables or bussing soiled dishes.

—Touching the food-contact surface of tableware—the tines of a fork instead of the handle, the rim of a cup instead of the handle—with the bare hands.

—Following the ancient culinary custom of tasting food with an extended index finger, or with the same spoon again and again. (Uninhibited Old World chefs—bless 'em—to the contrary notwithstanding.)

Smoking, if we may belabor the point, can endanger the health of both the worker and the customer. It is virtually impossible to smoke without exposing the fingers to droplets of saliva from the mouth. However small and unnoticed, these droplets may contain thousands of bacteria which can contaminate anything that fingers touch. Smoking, moreover, is a double hazard. The contamination route can operate in the reverse: organisms may pass from a soiled object to the hands, to the cigarette, to the lips and mouth.

To bring our "endless" list to a practical end, let's remember one simple and easy rule: *Hands that have touched contamination must never touch food—without hand-washing.*

Caution—Hazardous Customers!

The customer is the one foodhandler you can't make safe. You welcome his patronage—in fact, can earnestly solicit it. You obviously can't select, train, or supervise him (except in the most extreme cases). Yet the patron is potentially a dangerous source of contamination. He enters the establishment wearing street clothes, handles food even though he may know he is carrying infection, and is free to run the full gamut of unsafe behavior—from unguarded coughing to inadequate handwashing. While the opportunity for him to introduce contaminants may be less significant in sit-down dining, it is ominously large in self-service arrangements. If food is exposed to the customer in, say, a cafeteria line or salad bar, you must somehow discreetly and effectively erect the sanitary barrier between the contagious customer and food served to other patrons.

Such food products as cold sandwiches should be wrapped. "Sneeze guards" (glass baffles covering exposed foods) should be used to protect steam tables and cold food displays. Each container of food in a self-service buffet should be supplied with its own long-handled tongs, fork, or spoon. Dishes should preferably be stored in automatic dispensers to prevent customers from touching more than one at a time. Cream and sugar are best kept in closed containers.

THE RULES IN ACTION

Setting up the numerous and complex rules necessary for safe foodhandling is one thing. Making it possible for the rules to be observed is another. And seeing to it that the rules *are* followed is still another. In putting the rules into action, the manager should make maximum use of three effective tools: adequate facilities, regular training and constant supervision.

Adequate Facilities

As in every aspect of a foodservice operation, safe foodhandling requires proper equipment—in this case, adequate washing and dressing-room facilities. The object is to encourage the worker to obey the rules by making it convenient and comfortable for him to do so.

He will need proper facilities from the moment he arrives on the job. To start with, he should have a dressing room where he can change into work clothes and clean up, and a locker for the safekeeping of clothing and personal effects. Conveniently located but separate from foodhandling areas, this room should be clean, well-lighted, and uncluttered. If it isn't—if it's dingy, cramped, or hard to reach—he will tend to avoid using it.

A rest or "break" area separate from the foodhandling areas is also desirable and, in fact, is required under the ordinances in some localities. Employee restrooms, equipped with self-closing doors, ought to be provided, separate from those for customers. Convenience factors have to be considered in locating the lounge area and restrooms just as they were for the dressing rooms, and basically for the same reason.

From the standpoint of sanitation, the most important employee facilities are those for cleaning up. No employee should be allowed to return to work after a break, meal, or visit to the restroom without washing his hands. Depending on his job, the worker may need to shower at the beginning and/or at the end of the day. He may also need to wash and change clothes at various times in the course of his shift.

Lavatories and showers must be kept scrupulously clean, and strict housekeeping should be demanded of those who use them. The ironic alternative is that a facility intended to promote sanitation may turn Frankenstein-like into a dungeon of disease. And, obviously, the facilities cannot serve their purpose unless they are well-equipped and continuously supplied with soap, towels, waste receptacles, etc.

Since it would be impossible for a worker to get through a day without exposing his hands to contamination, readily available handwashing stations are a must. Convenience, again, is an important factor. If the lavatory is too hard to get to, workers will be tempted to do the forbidden thing—use a sink in which food or utensils are cleaned—raising the spectre of cross-contamination.

Ideally, the hand-washing station ought to have faucets that can be actuated with the foot, knee, or elbow. More economical self-closing, hand-operated faucets are satisfactory in smaller operations in which the lavatory will not be used very frequently. In either case, the hand-washing facility should supply hot and cold water through a mixing faucet. Many operators find it best to use premixed water at between 110°F (43°C) and 120°F (49°C). If the water is uncomfortably hot or cold, workers will shy away from it.

For hand drying, management may supply disposable paper towels, continuous cloth towels dispensed from a self-retracting cabinet, or forced-air blowers. Some authorities consider the blowers to be the most sanitary, but workers are inclined to think they are too slow. If paper towels are used, disposal hampers must be provided. Hand-drying facilities should also be available at food preparation and utensil sinks—but, of course, not as a follow-up to handscrubbing. The object is to discourage workers from using aprons or wiping cloths to *dry* their hands.

Training The Safe Foodhandler

Let's say we have established the policies for safe foodhandling, and provided the facilities so that strict observance is physically possible. Now to develop the right attitudes among workers so they will obey the rules willingly and consistently. This calls for a regular program of instruction and indoctrination. A one-time lecture won't do. Training must be a continuing process.

As we have noted, training should begin at the very moment of hiring. The new worker must clearly know which practices are unsafe and not to be tolerated, and which are desirable and encouraged. And he must know the reasons behind these policies.

Training should not be a tedious exercise and need not be. We are, to say the least, dealing with subjects that are not without elements of excitement and challenge: the staggering complexity of micro-organisms that inhabit our bodies and the world around us, and the intricate maneuvers which we, of the food-service team, employ to protect food, and our livelihood, from their assaults.

Supervising Employees

No program of personal hygiene can succeed without constant supervision. In the heat of action, at peak serving periods, even the most conscientious employee may forget. He may sweep an errant forelock from his perspiring brow with a bare hand. He may disjoint raw poultry and then, without hand-washing, slice some baked ham. He may neglect to trade his heavily soiled shirt for a clean one. Supervisors must always be on the alert to spot unsafe practices that may creep back into an employee's work.

One excellent approach to foodhandler supervision is to apply the same health standards to regular workers as are used in screening prospective employees. In a discreet manner, check employees daily for infected cuts, burns, boils, respiratory symptoms, and other evidence of infection that can be transmitted through food.

Under the law in many localities, the proprietor who knows or suspects that an employee has a contagious disease, or is a carrier, must notify health authorities immediately. Depending on the disease, the employee may continue to work after medical treatment. (He may, for example, be required to wear rubber gloves because of a skin infection, or be assigned to another job which doesn't involve direct contact with food.) More serious ailments may require treatment which excludes the worker from the establishment for a considerable period of time. In extreme cases, a worker may be ineligible for foodservice employment indefinitely.

As in other management positions requiring continuous surveillance of the operation, monitoring a safe foodservice means developing good supervision routines.

It should become second nature for the manager to run through a mental check list as he makes his daily rounds:

1. Are workers wearing clean uniforms?
2. Are they free of body odors?
3. Is the clean-hands policy being strictly observed?
4. Are workers wearing hats, hairnets or other hair restraints?
5. Are any observed fingering a pimple or scratching at the head and face?
6. Do they smoke or eat in food preparation or serving areas?

7. Are their fingernails short and clean?
8. Do they spit in unauthorized places —in sinks, on the floor, in a disposal area?
9. Do they cough or sneeze in their hands?
10. Are they wearing ornate rings, dangling bracelets, wristwatches, or other easily soiled personal decorations while handling food?
11. Do they use wiping cloths to remove perspiration from their faces?

Ideally, the employee should feel free to report that he has an infection. He will be more likely to do that if he knows he will not be penalized with loss of work and pay. If he has a slight cold or a minor infected cut, he may well be given a temporary job not involving food handling. There is no pat answer to these problems, but the manager must strive to maintain a climate in which no member of his staff will be discouraged from coming forward with information about dangers to health.

Supervising Yourself

While you, the boss, are watching your employees, you should also be watching yourself. One of the most fruitful tools of management is to set a good example. The cook who sees the manager walking through a food preparation area with a cigarette in his mouth probably thinks, "If the boss can do it, why can't I?"

The leader who doesn't practice what he preaches will not lead effectively, and he can't lead in the desired direction unless he himself knows the way. Some managers have never been trained in sanitation or have been trained but do not appreciate its importance.

It is natural for a person to think that he could never be the source of a sanitation problem. The supervisor is probably the most experienced person in the operation he manages—and unfortunately he may be most immune to learning. Even if he doesn't think he already knows it all, he may consider he is too busy to participate in training sessions.

A manager confronted with a reluctant supervisor may try a subtle technique to change his attitude. He may ask him to make a sanitary survey of another department or operation. Unsafe practices observed by the supervisor in such an inspection will very likely suggest possible defects in his own department.

IN SUMMARY

Most food contamination is caused by foodhandlers with skin infections and diseases of the respiratory system or intestinal tract.

Micro-organisms are commonly transmitted from the body to the food product by unclean hands, as well as by coughing or sneezing.

In most localities, the foodservice manager is required by law to prevent individuals with a communicable disease from working in contact with food, and to report diseased employees to the regulatory agency.

The manager must build a sanitary barrier to protect his customers from contaminated food. He can do this by carefully screening out diseased individuals at the hiring stage, establishing clearly understood policies against unsafe personal habits, and requiring personal cleanliness of foodhandlers.

Thorough handwashing after contact with contaminated objects is an indispensable rule of personal hygiene.

To be sure that his sanitation policies are observed, the manager must provide adequate personnel facilities, regular training, and constant supervision.

A CASE IN POINT

The pantry girl had a problem — dishpan hands. She didn't call it that, but the manager got the message.

"Your lecture on hand washing really got to me, Mr. M, and I've been doing what you said. But it takes so much time — and my hands are raw!"

Tim was glad to see his training efforts taking effect but this report set him wondering. He thanked the employee for her interest, ordered some hand lotion for the locker rooms and gave the matter some thought. He noticed that this young lady was forever leaving the kitchen between periods at the vegetable preparation sink. Why wash your hands, he asked himself, when they are as clean as the lettuce people are going to eat? The practical answer was simple enough. Paper towels. But what was the principle involved? That night he reviewed his sanitation training notes and came up with an idea: He would define two basic conditions, two *modes,* for a foodhandler's hands: (1) The *foodhandling mode,* when the hands are sanitary and may only need drying with a clean towel, and (2) the *food-hazardous mode,* when a sanitizing scrub or washing is in order.

Tim figured that if his people got into this habit of thinking they would instinctively do the right thing, and not over-do it.

Do we accept his approach to the problem?

Yes. Very good, with the proviso that, whenever cleaning agents or solvents other than water are used, the washing-up will be done well clear of food preparation and serving. That covers the back-of-the-house. For the waitress out front who must handle soiled dishes, wipe-cloths, money, etc., between customers, there is no substitute for a trip to the wash basin.

MORE ON THE SUBJECT

For further reading, the student is referred to the following sources described in the Bibliography, Appendix B.

Reference 4 *Quantity Food Sanitation* (Longrée, 1968), Chap. V-VIII.
           5 *Sanitary Techniques in Food Service* (Longrée/Blaker, 1971), Part III.
           7 *A Self-Inspection Program for Foodservice Operators* (NRA, 1973), Check Sheet 1, "Personal Safeness."
          10 *The Sub-Standard Washroom* (NRA, 1966).
          12 "48 Ways to Foil Food Infections" (Bete, 1970).
          23 *Food Service Sanitation Manual* (USPHS, 1962), Part V Sec. D.
          26 "Emerging Foodborne Diseases" (Bryan, 1972).

CHAPTER 5

Food Procurement and Storage

To paraphrase an old saying, it would be easier to make a silk purse from a sow's ear than to prepare a wholesome meal with contaminated or deteriorated ingredients. It is essential that completely safe materials be used in a foodservice. This means that food supplies must be in excellent condition when they arrive in the receiving area, and must be kept in that condition while awaiting preparation and service.

In previous chapters we have explored the fundamental role of micro-organisms in food contamination, and discussed the human factors. Now we are prepared to consider the importance of safe purchasing and storage practices in protecting food.

Government inspections help prevent the distribution of unsafe products, but the final responsibility rests with the manager — for several reasons. First, "Big Brother" is not able to inspect all food production, nor can he check the products at every point in the distribution network. In recent years quantity production and new products — pre-cooked and pre-portioned foods, for example — have made possible greater efficiency and economy in foodservice operations, but along with these developments have come new hazards to the consumer. The additional processing steps and longer lines of supply have increased the number of events which could cause contamination and spoilage. For example, a frozen food product which was perfectly safe when it left the manufacturer's plant may be damaged by thawing in the railroad car or truck that carried it to the wholesaler, and again in the van that carried it to the point of use. Within the foodservice establishment, the manager is of course obligated to guard the sanitary quality of food products in storage.

In this chapter, each of the major categories of food will be analyzed as to:

- **Sanitation criteria for selection and purchase**

- **Inspection procedures on receipt**

- **Safe storage requirements**

GOVERNMENTAL CONTROL

Government inspection programs now cover meat, dairy products, shellfish, poultry, and eggs — in the production and processing stages. In addition, governmental agencies inspect canned and frozen food processing plants. Various federal agencies have established regulations and standards for foods shipped in interstate commerce. Many state and local agencies have, in turn, adopted these criteria (or imposed even more stringent ones), and monitor production and processing.

The supervisor responsible for procurement should have an understanding of the extent of these inspection services and their significance. He may consult his local health officer for information on state and municipal food regulations. Two departments of the federal government, in the main, have cognizance over food protection standards of direct significance to foodservice establishments: the Department of Health, Education and Welfare (HEW) and the Department of Agriculture (USDA). Within HEW, the Food and Drug Administration and other agencies of the U. S. Public Health Service exercise inspection control over food processing and provide guidance for state and local regulation of foodservice operations. As will be seen from the following, the USDA is largely engaged in the grading and inspection of meat, poultry and dairy products.

MEAT AND POULTRY

Meat, meat products, and poultry shipped in interstate commerce are graded and inspected by the USDA. To help guarantee that products are wholesome, unadulterated, and sanitary, it is advisable to buy only government-inspected meats and poultry. Look for the inspection stamp and shield-shaped grade symbol on meat carcasses. Poultry should be checked for the USDA tag (usually located on the wing) which indicates that the product has been inspected and graded.

MILK AND DAIRY PRODUCTS

The milk surveillance program in the United States is one of the most successful food protection programs ever instituted. As a result, there have been almost no foodborne outbreaks attributed to dairy products in recent years. The program stemmed from early recognition that milk and milk products were a major potential source of foodborne illness. It was also realized that standards of sanitation had a direct bearing on the keeping quality of milk.

Virtually all dairy production areas are now covered by USDA certified inspectors. They check the health of dairy animals as well as the sanitary manner in which the products are processed and transported.

SEAFOOD

On a voluntary basis, fish and shellfish are certified as to wholesomeness by two federal agencies. The U. S. Department of the Interior certifies fish which meet its standards of sanitary sources and processing, but most fresh fish currently being sold is not covered. Purchasers should depend on the advice of a reliable local dealer. Both fresh and frozen shellfish are certified by the U. S. Public

The Hazardous Trail from Farm to Table

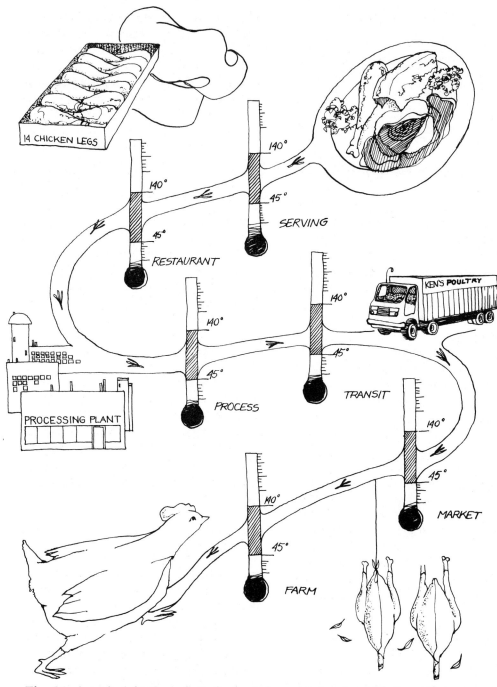

Fig. 14 A typical food product, fresh or frozen, runs the risk of passing many times through the temperature danger zone.

Health Service. Most states participate in this program, which covers the safeness of the water from which the shellfish are taken, as well as their handling, processing, and transportation. A list of approved shellfish sources can be obtained from the Public Health Service or the local regulatory agency.

INSPECTION AND STORAGE

A foodservice manager remodeling a plush dining room wouldn't accept tables or chairs that were scratched, chipped or broken. He should be even more particular about the food products coming into his establishment.

Inspection of food products at time of receipt is critically important. Even governmentally approved foods may be rendered unsafe by mishandling in transit. Once in the establishment, improper storage can of course damage their sanitary or culinary quality.

A well organized receiving system should include inspection as well as inventory control and billing. Personnel involved should be able to judge quality, check temperatures of frozen foods, detect damage, and spot insect infestation.

(As a general rule, pre-cooked food, held-over food and other ready-to-serve menu items should be stored in their own refrigeration unit whenever separate facilities are available. Without exception they must be enclosed in sealed containers to exclude raw food contaminants. This isolation will also serve to prevent loss of aroma and intermingling of food flavors.)

Most perishable food items demand attention the moment they are received.

The wise manager will arrange with suppliers to make deliveries during off-peak hours, and plan ahead to make sure that sufficient refrigerator/freezer space is available. In some well equipped operations, there is a refrigerator and freezer in the receiving area for quick and temporary storage. Many will also provide for preliminary washing of produce to prevent insects and excessive soil from being brought into the house. If the foodservice establishment is indeed a fortress against enemy vermin and microorganisms, its receiving area should be the first line of defense.

No matter how meticulous you may be as a manager, you must to a large extent still depend on your supplier for quality foods. But you should reassure yourself that he is indeed dependable. Have a look at his facility. In the course of a visit, ask yourself: What are his standards of sanitation? Does he follow approved practices in handling the product? If he handles refrigerated or frozen foods, is he equipped to do the job? Does he have enough warehouse space? Are his delivery trucks adequately refrigerated? Is he willing to make deliveries when your personnel are not rushed with other duties?

FROZEN AND REFRIGERATED FOODS

INSPECTION. Visual inspection of frozen food may be deceiving because of the possibility of some rather exotic microbic phenomena. Freezing retards the development of most bacteria, but some bacteria, the psychrophiles, flourish at temperatures as low as 19°F. Harmful organisms in this category may therefore gain an advantage over harmless organisms at freezing temperatures and compete unequally when the temperature rises. Certain cold-loving bacteria also produce enzymes on thawing which will continue to break down foods even after refreezing. The result is that foods which appear solidly frozen may have undergone some progression of disease or spoilage agents.

On receipt, frozen food should be checked for signs that thawing has occurred at any point between original processing and delivery. Obvious signs of thawing include fluid or frozen liquids in the food carton. Large ice crystals in the product itself indicate that it has thawed and then refrozen.

STORAGE. The following guidelines were drawn from the code of recommended handling practices issued by the Frozen Food Industry Coordinating Committee (Sept., 1970):

1. Freezers should be maintained at an air temperature of 0°F (about −18°C) or lower.

2. Products should be placed in storage facilities of approved design promptly after delivery, and removed from storage only in quantities that can be used immediately.

3. Frozen food inventories should be rotated on a "first-in, first-out" basis.

The rule on immediate storage of frozen foods does not strictly apply if the delivered items are going to be used soon. The amount of time meant by "soon" will depend on the type of food, but in every case the products should be kept refrigerated, and not held at room temperature. For frozen vegetables and pre-portioned meats, "soon" means right away; for average-sized roasts, it is the next day; and for turkeys, two days.

Refrigerators and freezers have their limitations, and are no substitute for good advance planning of menus and food purchases. If you make an error and buy too much food for your needs, it won't do merely to store the surplus until some distant date when it can be used up. Lengthy storage means increased opportunities for contamination and spoilage.

While proper freezing can halt the growth of bacteria, it certainly will not destroy all of them. Some micro-organisms multiply at temperatures as low as 19°F (about −7°C). Freezing and refrigeration also can't be expected to improve the culinary quality of foods; with some products, the reverse may be true. Among items which may deteriorate in freezer storage are hamburger, fatty fish (mackerel, salmon, swordfish), turkey, pork, creamed foods, custards, gravies, sauces, and puddings.

Strict maintenance of 0°F as the maximum temperature in any part of a freezer compartment is mandatory; even slight variations above that mark can be destructive. It has been found, for example, that food deteriorates several times faster at 15°F (about −9°C) than at zero. Defrost damage is cumulative. Each incident of thawing adds to the damage, and cannot be corrected by refreezing.

All of these factors impose a responsibility on the manager to make certain that his freezer is actually maintaining the required temperature. Accurate thermometers should be kept in easy-to-check locations.

Much the same precautions apply to refrigerator units. An air temperature of 40°F (about 4°C) or lower should be maintained, and regularly checked with a reliable thermometer.

Care should be taken not to disturb the actual temperature through storage of large amounts of warmer foods. Most commercial freezers and refrigerators are designed to keep cold foods cold. They may not have the capacity to bring warmer food items down to the proper temperature quickly enough to avoid possible damage. Small quantities of warm, or even hot food can, however, be accommodated without harm. Some freezers are specifically equipped to receive thawed products and bring their temperature to 0°F, but this capability is not typical of conventional storage units.

Too much traffic, allowing warm kitchen air to flood the interior of a refrigeration space, can also affect temperature unfavorably; keep open-door periods to a minimum. The time to open a refrigerator is after you have decided on the products you want to break out.

Ideally, use separate freezers or refrigerators for each main food category. In this way, it is possible to provide the optimum temperature and humidity required by each type. It also helps to prevent odor-absorbent foods such as dairy products from picking up strong odors as from fish, and from some fruits and vegetables. Large operations may have separate refrigerator storage for meat, dairy products, fruits and vegetables, and fish.

Where this flexibility doesn't exist, meats and dairy products should be stored in the coldest parts of the unit. And dairy items in particular should be tightly covered to prevent odor absorption. When it is not feasible to store ready-to-serve food separately, prepared foods ought to be stored above, not below, raw foods to prevent cross-contamination.

It is also desirable for foods stored in a freezer to be wrapped in moisture-proof materials, and labelled with the date and product description to facilitate first-in, first-out practice. Food cartons should not be stored in such a way as to interfere with the circulation of cold air. For the same reason, shelving should be of the open type. Lining the shelves may make them look neater, but it cuts refrigeration efficiency.

In walk-in coolers, foods should be stored away from walls — to prevent insect or rodent nesting — and off the floors to avoid contamination during cleaning.

Proper sanitation does not stop at the refrigeration door. Freezers should be defrosted and cleaned periodically. Refrigerators ought to be checked regularly for the presence of debris (which blocks air flow), mold, and objectionable odors. These are all signs of poor storage or inadequate cleaning.

Ease of cleaning should be a consideration in purchasing a new refrigeration unit. Look for interiors that are free of corners and sharp edges. Surfaces should resist corrosion, chipping and cracking. Loose particles may end up in the stored food. Parts that are easily soiled — the shelves, for example — ought to be removable so they can be thoroughly cleaned.

CANNED PRODUCTS

INSPECTION. As with other foods, the two quickest and simplest tools to use in checking the acceptability of canned foods are your own senses of sight and smell. Canned goods that appear abnormal in odor, color or texture should be promptly discarded without even taste-testing. In general, canned goods must be rejected if they have:

1. Pinholes. To check for these tiny holes caused by the action of food acids during prolonged storage, remove the contents and rinse the can. Then hold it up to a strong light and examine the interior for pinholes.

2. Penetrating Rust and Dents. If the rust can be rubbed off the can exterior, or if the rust or dents do not penetrate to the interior, the contents will normally be usable.

3. Swells (or Swellers). Both ends of the can bulge outward because of gas produced by bacterial action. The ends do not yield when they are pressed with a finger.

4. Springers. One or both ends bulge outward because of bacterially caused gas. In this case, however, the end will yield to finger pressure and spring back to the bulged condition on release.

5. Flippers. Both ends are flat, but one end will bulge outward when the other end is pressed. This condition is caused by bacterial or chemical action which produces gas.

STORAGE. Canned foods should be kept in a dry area at a temperature of 50°F to 70°F (10°C to about 21°C). Higher temperatures are likely to accelerate bacterial action and food deterioration. For this reason, cans should not be exposed to sunlight or stored near heating pipes and vents.

DRY FOODS

INSPECTION. Cartons of cereals, sugar, dried fruits and vegetables, and flour should be dry and undamaged. Punctures, tears, or slashes in the package may indicate insect or rodent entry. If the outside of a container is damp or moldy, this condition may extend to the contents, raising the possibility of advanced microbic contamination. Most dry foods are poor media for bacteria, but a touch of moisture can radically change this.

Dry foods that are off-color or have an unusual odor should also be rejected. Spoiled baked goods, for example, may have a sodden or slimy appearance.

STORAGE. For maximum shelf life, dry foods should be kept in a dry place that is clean, well ventilated, relatively cool, and secure against insect and rodent invasion. Temperatures of 50° to 70°F and a relative humidity of 50 to 60% are best. Windows in the storage area should be blocked off with shades to prevent sunlight from raising the air temperature or affecting food quality. Exposure to bright light can change the colors of spices and chocolate, and help turn cooking oils and fats rancid.

While dry items are not usually subject to bacterial action, they are prime targets of insects and rodents. It is considerably easier to prevent these pests from gaining entrance than to eliminate them once they have become established. Doorways to the storage area should be closed off with solid or screened doors to stop flying insects. Cracks and crevices in the floor or walls should be filled in so they won't harbor dirt or vermin, or allow the entry of rodents. Store foods off the floor and away from walls to eliminate these hiding places for pests. Make it a habit to inspect storage areas regularly for insect and rodent infestation, as well as for signs of other damage or spoilage.

Steam pipes, ventilation ducts, water lines and other conduits have no place in a well-designed storeroom. Dripping condensation or leaks in overhead pipes can promote bacterial growth in such normally stable items as crackers, flour, and baking powder. Leaking overhead sanitary lines are obviously a highly dangerous source of contamination for any food. If conduits can't be avoided, store foods as far away as possible and rotate them frequently. To guard against rodents, make sure there are no openings where pipes pass through the walls.

Food items should be stored in their original containers on pallets or shelves. A minimum six-inch clearance above the floor is desirable to permit cleaning and to facilitate air circulation. Items in damaged containers should be used immediately, assuming they are otherwise acceptable. If there is any doubt about the fitness of the contents, the food should be inspected by a health department representative.

A food storeroom is for food only. Foodservice operations commonly use a number of substances that are poisonous to humans, including insecticides, cleaning compounds, disinfectants, and lubricating oil. All of these non-foods should be stored in a separate, latched cabinet away from food areas. In addition to their potential danger, many of these non-foods broadcast strong odors which can be absorbed by nearby foods. The properly managed storeroom is off limits to smokers for the same reason — to

say nothing of the fire hazard.

The ideal storeroom is easy to keep clean and pest-free. Its floors are painted concrete, cement mixed with granite, or covered with quarry tiles, and have a drain to dispose of wash-up fluids. (Wood floors are too difficult to clean and to keep vermin-proof; asphalt tile floors may crack under heavy loads, opening up hiding places for insects.) Its walls are covered with washable paint or glazed tile. Shelving, pallets, and tabletops are of non-corrosive metals. Bins for flour, cereals, grains, and dried vegetables are also of non-corrosive metals, and covered to keep out moisture and vermin. Easy to clean or not, the storeroom should be regularly swept and swabbed down to keep it free of contaminating dust and debris.

MEATS

INSPECTION. In the unfrozen state, meat is normally firm and elastic to the touch. It is also free from slime deposits, abnormal odors, and discoloration. Generally, the first indication of deterioration will be the appearance of slime, caused by microbic growth on meats stored under conditions of high temperature and humidity.

Aged beef may have a somewhat sour odor. Fully decomposed meat produces an unmistakable stench. Hamburger is the most likely form of meat to become putrid, especially when it has been frozen and then defrosted by unsafe methods. But it can become rank even under refrigerated storage.

Since pork spoils first inside the tissues near the bone (beef spoils first on the surface), an abnormal odor may not be readily detected. Test for over-age pork by pushing a knife into the flesh, and then sniff the knife tip.

Deteriorating meats display a wide variety of abnormal colors. Brown, green, or purple blotches are all signs of a microbial attack. Black, white, and green spots indicate molds. Unless the growth is extensive, the meat can be used after trimming.

To be acceptable, meats should have the appearance indicated in the following guide:

Beef ....A bright, cherry red for fresh cuts; aging darkens its color.

Veal ....Typical carcasses are grayish-pink; lean portions have a smooth, velvety texture.

Pork ....Fat portions are white and firm; meat is light pink. Deterioration is usually evidenced by darkening of lean meat and discoloration and rancidity of rind.

Sausage ....Free of slime and mold, which indicate decomposition. External mold is common on dry sausages such as salami, but it can be washed off and is considered harmless if it is confined to the casing. Mold penetrating a broken casing may impart a moldy taste, but is not considered injurious to health.

STORAGE. Place fresh meats in refrigerator storage as soon as delivered.

All types require a temperature of 30°F to 36°F (about −1°C to 2°C), though some meats can tolerate longer storage than others. The maximum recommended storage time for beef is one month; for mutton, lamb and pork, less than a month; for veal, a few days.

POULTRY

INSPECTION. There are numerous indications of spoiled or inferior poultry, almost all readily observable. Soft, flabby flesh and sunken eyes usually mean an inferior product. A purplish or greenish over-all cast, and a greenish discoloration around the neck and vent may mean staleness, improper bleeding after killing, or improper handling. Other signs of spoilage: an abnormal odor, stickiness under the wings and around joints, and darkened wing tips.

STORAGE. Use in less than a week after delivery, and store at 30°F to 26°F, ideally in a relative humidity of 75 to 85%.

FISH AND SHELLFISH

INSPECTION. Few food items suffer from improper storage as much as fresh and frozen fish. Do not use fish that has thawed and been refrozen before it reached your establishment. Refrozen fish will have soft, flabby flesh, a sour odor, and an off-color. The paper in which it is wrapped will be moist, slimy and discolored. The bottom of the shipping carton will have ice formed in the refreezing position, and the container itself may be deformed by internal pressure. Frozen fish fillets of poor quality will also be brown at the edges.

Fresh fish can be easily distinguished

from stale fish by marked differences in appearance. Fresh fish have bright red, moist gills. The eyes are bulging and clear. The flesh and the belly area are firm and elastic. Fresh fish also do not have a noticeably strong odor.

An unacceptably stale fish presents a complete contrast. The gill slits are gray or gray-green, and dry. The eyes are cloudy and red-bordered. The scales may be loose and show deposits of slime. The flesh is soft and yielding. Apply finger pressure and the impression will remain. If fish or shellfish have an ammonia odor, the deterioration is advanced and the products should be immediately rejected.

In addition to these surface indications, unacceptable fish may exhibit internal signs of decay. With freshwater varieties, look also for parasites and disease — tumors, abscesses and cysts.

Fresh lobster and other shellfish should be alive on delivery. The shells of clams and oysters will be closed in the living state, partly open if dead. The edible portions of frozen lobster should have firm flesh and a normal odor. They are acceptable for use so long as there is no evidence of shrinkage or a change in normal contours.

STORAGE. If refrigerated at 32°F (0°C) and iced, fresh fish may be stored up to three days. If kept at that temperature but not iced, the fish should be used in 24 hours.

EGGS AND DAIRY PRODUCTS

INSPECTION. Fresh eggs should not be more than 30 days old when received from the vendor. The shells should not be cracked, because of the danger of invasion by salmonellae. To test the freshness of a shipment of eggs, break one open. If the albumin clings to the yolk, and the yolk itself doesn't break too easily, the egg is acceptable. Frozen or dehydrated eggs must have been pasteurized.

Milk does not need to be marked "Grade A" to be usable, but the cartons certainly should carry the pasteurization label. Always look for it. Unpasteurized milk is a potential carrier of a wide variety of diseases, ranging from salmonellosis and shigellosis to undulant fever and diphtheria. The pasteurization process — heating milk to the right temperature (138°F minimum) for the right amount of time — destroys significant pathogenic organisms with minimal chemical change.

Generally, milk intended for use as a beverage must be packaged in individual containers. In some localities, regulations permit the serving of beverage milk from refrigerated dispensers. Dried milk or milk in bulk containers (5 to 10 gallons) may be used for cooking purposes.

Milk with an off flavor may still be wholesome, but routinely it should be rejected. It is also not acceptable if its temperature was above 40° (about 4°C) on delivery.

Butter should have a sweet, fresh flavor, uniform color and firm texture. It should be received in clean, unbroken containers, free of specks or other foreign substances.

Cheese ought to be checked to see that it has characteristic flavor and texture, as well as uniform color. If the cheese has a rind, the rind should be clean and unbroken.

STORAGE. Most dairy foods readily absorb strong odors in the vicinity, including flavors from other foods. For this reason, dairy products ought to be kept tightly covered, and stored away from strong odor sources. A storage temperature below 40°F is necessary, and a humidity level of 80 to 85% is desirable. Milk, cream, and cottage cheese should be used within three days after receipt. After a container of dried eggs has been opened, it should be tightly covered and refrigerated for storage. Cheese may be stored almost indefinitely, but any moldy portions should be removed and discarded before serving.

FRUITS AND VEGETABLES

INSPECTION. For fruits, taste is the best test of quality. Many housewives rely on appearance as an indication of quality, but this criterion may not be in all respects dependable. Blemishes can be present even when the flavor quality are unimpaired. As a rule, abnormally large specimens of fruit are more woody and coarse in texture than smaller ones.

Fresh vegetables must be handled with extreme care because of their perishability. Pinching, squeezing, or unnecessary handling on receipt will bruise them, leading to decay and premature spoilage.

Fruits and vegetables show spoilage in a number of ways. Bacterial breakdown of onions, carrots and celery is revealed through a waterlogged, mushy look that is sometimes accompanied by bad odor. Citrus fruits are subject to a blue mold rot; leathery brown spots indicate deterioration of lemons and strawberries. Light brown or tan spots result from deep decay in peppers and eggplant. A gray mold attack strawberries, beets, endives, plums and cherries.

The familiar reddish brown areas on lettuce are caused by washing in contaminated water, which in turn puts bacteria to work.

STORAGE. A temperature range of 35° to 45°F (about 2°C to 7°C) and a humidity range of 85 to 95% are preferable for almost all fruits and vegetables. A rare exception is the banana, a tropical fruit sensitive to low temperature. Citrus fruits may be held under refrigeration up to two weeks.

IN SUMMARY

The foodservice manager is assisted in obtaining safe food products by various government inspection and grading programs. He nevertheless has the final responsibility to inspect delivered items, and to store them promptly and properly.

Each major category of food displays characteristics which mark it as fresh and acceptable, or decayed and unacceptable.

Once the manager has accepted food items from the supplier, he must safeguard them until preparation and serving time. To do so responsibly requires an understanding of the proper uses of freezers and refrigerators, as well as a carefully designed dry food storage area.

Isolation of raw foodstuffs from prepared menu items is an important factor in the storage of food to avoid cross-contamination and cross-flavoring effects.

In planning and operating storage facilities, the manager must make certain that required temperatures are maintained, that standards of cleanliness are observed, and that every effort is made to prevent contamination by vermin and rodents.

A CASE IN POINT

What a lucky break! Lingering over his morning coffee, Tim thought to himself there was no telling how far that banged-up can of beans might have gotten if he hadn't spotted it.

Just thinking about it made him angry all over again. "What do you mean sending me bad stuff?" he'd shouted at his supplier over the phone. "I'm sorry . . . it happens," was the apologetic answer.

During the breakfast rush Tim would ordinarily be indoors — in the kitchen or dining room. But this morning, he had walked out to the receiving dock to take a breather. It was then that he saw the carton of canned beans being unloaded from a truck, and noticed the big hole in one side of the box.

He ripped open the carton, extracted the can that showed the worst damage, and flashed it to his workers. "Watch it!" he warned them. "This can be poison when it gets like this." He opened the can, discarded the contents and held it up to the sky. Sure enough, tiny spots of light showed through the metal.

That was all he needed to see. Tim resolved to do two things to make certain there would be no repetition. First, he would get a new supplier, even though his present one was an established firm and had given him good service. Second, he was going to spend more time on the receiving dock, checking up on incoming goods.

Has Tim gone too far, or hasn't he gone far enough?

Tim's intentions are the best, but very likely he has overreacted. Certainly he should select reliable suppliers and expect them to provide safe and wholesome products. But accidents do happen. The foodservice manager has to make sure that incoming foods are properly inspected and stored, of course, but he can't be everywhere. Instead of spending more time on the receiving dock, he would do better to delegate that responsibility and then train the assigned worker to distinguish between acceptable and unacceptable supplies.

MORE ON THE SUBJECT

For further reading the student is referred to the following sources described in the Bibliography, Appendix B.

Reference 7 *A Self-Inspection Program for Foodservice Operators* (NRA, 1973).

23 *Food Service Sanitation Manual* (USPHS, 1962), Part V, Sec. B.

25 *Protecting Our Food* (USDA, 1966), "Storage and Warehousing."

33 *Food Service Operations* (USN, 1971), Chap. I.

34 *Code of Recommended Practices for the Handling of Frozen Food* (FFCC, 1970).

CHAPTER **6**

Protecting Food in Preparation and Service

"I sometimes think foodhandlers are catastrophes looking for a place to happen," a veteran restaurateur once remarked, "and a fair number of them seem to be heading for my kitchen!"

Our anxious friend makes his point, if somewhat dramatically, but the prospect is not really so grim. The path to a safe foodservice may seem to be strewn with boobytraps but it is fairly well lighted and there are markers along the way for those who will read them.

If, as some literary analysts say, the basic plot of every story puts the hero in a hole and then proceeds to get him out of it, our story has just about arrived at the rescue phase. The previous four chapters have described a rather thoroughgoing predicament for the foodservice manager. It is now time to show our protagonist where he is on the map and head him out of the woods. His ultimate goal is to protect people from foodborne illness. He does this by protecting the food they eat. As we have said, food protection requires *preven-*

tive measures that seek to keep disease agents from getting into food, and *corrective* measures that destroy, or stop the multiplication of, those that do get in. The success of these measures depends on two fundamental operating principles:

- **Affirmative action to guard against contamination**

- **Strict time-and-temperature control to prevent progressive contamination**

At no point in the foodservice operation are we more intensively concerned with sanitation than during the climactic kitchen phase. The scene is set for the main action, the actors are on stage, and the important right — or wrong — things will now begin to happen in quick succession.

As we have seen from a previous discussion, the health and personal habits of the foodhandler are key elements in the drama. As the curtain rises for the main act, let's assume that our players are in "normal" good health, with no infections that could contaminate food beyond the redemptive powers of proper cooking, and that their personal hygiene is acceptable or better. How their food-handling practices stack up, we shall see in the examples that follow.

Let's assume also that the foodstuffs with which they are working have been procured from approved sources, were properly received and stored, and are in good-to-excellent condition. We will further assume that the stage director is a forehanded manager who from the beginning has followed the script for a safe foodservice, adhering to the familiar rule: "Start with the menu."

He has selected a menu compatible with his facilities, has trained his staff not to alter the menu in any way that would compromise food safeness, and has designed the menu to match anticipated customer demands. He has, in short, tried to head off all the problems associated with the flow of food, from menu planning to serving the customer.

THE 'DANGER ZONE' ON THE TEMPERATURE SCALE

From the moment the food arrives in the kitchen a red thread of danger weaves an intricate pattern through the fabric of our drama. The scale on a real-life thermometer will hardly be colored red between the upper and lower limits of safe food-holding temperatures, but the foodservice manager will do well to develop such a mental image as a reminder of the cardinal rule: "Food must not be allowed to stand within the temperature danger zone — 45° to 140°F (about 7°-60°C)."

If, as we have noted, man is the single most common source of food contamination, temperature is easily the most important factor in controlling contamination once it has occurred. It is manifestly true that certain foods have to be cooked to destroy natural and man-carried contaminants. And it is fairly obvious that cooking will serve also to counteract cross-contamination and re-contamination that may occur during extensive preparation and handling. What is *not* readily apparent is the danger in allowing already cooked food to dwell in the danger zone for prolonged periods as it sits on the steam table, in a double-boiler, bain-marie, chafing dish, or the like.

The temptation here is to keep the food hot enough for serving, but not hot enough to kill bacteria and neutralize their toxins. A delicate problem arises when, for example, the chef says to hold the prime ribs at a warming temperature below 120°F to keep them rare, or when he directs that the soup be kept well below boiling to preserve its flavor. Any such conflict between culinary and sanitary quality considerations must always be resolved in favor of safeness, but this can mean lost time and wasted food if re-refrigeration and re-heating or premature disposal is involved. The answer lies in good timing and, as it so often seems, more than a little good luck.

Temperature control is a straightforward concept where most food items are concerned, but it appears to be not very well understood as it applies to frozen foods. There are a number of ways to account for this but perhaps the clearest explanation derives from the simple fact that they *are* frozen and may require an extra step in preparation. To defrost or not to defrost — the question arises again and again. Many frozen foods, particularly frozen vegetables, are pre-portioned and packaged for cooking without thawing beforehand. Yet it is not uncommon to see them set out in a foodservice kitchen to defrost at room temperature, inviting progressive contamination from bacterial growth. Unless frozen vegetables need further cleaning or additional preparation — slicing, soaking, seasoning, mixing with other ingredients — they may with rare exceptions be cooked in the frozen state. Directions from the processor will usually indicate how they are to be handled, including cooking time and variation in cooking time if they are to be transferred directly from the freezer to the pot. The procedure for thawing vegetables that are to receive further preparation is uncomplicated. Gradual defrosting at chill-box temperatures is preferable for retention of texture and flavor. Defrosting in cold water, with the product shielded against moisture, is acceptable, as are similar accelerated methods.

Other frozen products such as meat and poultry will often need to be thawed to permit further cleaning and preparation: a whole chicken to be sectioned for frying, a turkey to be stuffed, a pork loin to be breaded, shrimp to be shelled and de-veined, lamb shoulder to be trimmed, skewered and marinated. Except for the case in which room-temperature thawing can be accomplished in a very short time, these items should be defrosted under refrigeration or with cold water. Since this process will often require many hours, good planning is very important.

The time-and-temperature principle applies in a special way when we consider the vulnerability of frozen foods that have been *pre-cooked*. And the potential problem associated with these "convenience" products is all the more significant in view of their increasing popularity. Pre-cooked foods are not categorically more hazardous than other food products, but may be more sensitive to mishandling because of additional excursions through the temperature danger zone in the pre-cooking process. They have, in a word, already been through an entire kitchen phase. The processor's instructions are usually quite explicit, and the foodservice manager will be well advised to demand strict compliance on the part of his staff.

Some practical illustrations of the time-temperature principle are presented in the three examples which follow.

The Untimely Turkey

We are preparing a 20-pound frozen tom turkey. The bird must be thawed prior to preparation because it requires additional cleaning and trimming. It would take two days to defrost him in the chill-box and the cook is in a hurry. He lays the gobbler out on a work table in the kitchen to thaw overnight. The air temperature is about 75°F. The skin and outer layers of flesh will thaw rather rapidly, but will slow down heat transfer to the inside of the bird. As a result, the inner parts will thaw very slowly, as the skin and outer layers rapidly reach temperatures favorable for the growth of micro-organisms. In an overnight thawing period these organisms will increase to enormous numbers.

Actually, several problems have developed because of this procedure:

— A veritable "hot bed" of bacteria has been produced which will contaminate everything in the vicinity: the worker's hands, the work table, the cutting board, knives, cutters and other utensils. This is a classic example of *cross-contamination.*

— Toxin-producing organisms present will generate extensive reservoirs of poison in surrounding tissue. If subsequent cleaning, cooking, and further preparation are not done with extreme care, a dangerous dose of disease agents may be served to the customer.

— Allowing the microbial population to run wild has diminished the culinary quality of the turkey, introducing rancid taste or other off-flavors.

The same thing can happen with almost any frozen product that is improperly thawed. Food preparers must work under the assumption that no raw product is totally free of contamination, and that safe thawing procedures are of primary importance. Here are some pointers on acceptable defrosting methods:

1. In a refrigerator: This is a slow, space-consuming method which requires advance planning, but it is probably the most desirable. Provide for a free flow of air to facilitate heat transfer.

2. In cold moving water: Water should be clean and its temperature maintained at or below 50°F (10°C). Warm or hot water thawing is not acceptable. The product can be protected from contaminants or undesirable wetting by the use of plastic film or other waterproof envelope.

3. Special equipment: Proprietary devices are available for thawing frozen products. Some of these units incorporate warming elements to remove the chill generated by frozen material, and all are designed for efficient air flow to maintain uniform defrosting temperature.

Assume now that our turkey has been left out overnight to thaw and is to be prepared for cooking. The chef will clean it, trim it, and take other steps to get it ready for the oven. As soon as he has done so, the possibility of new contamination problems arises:

— Heat transfer may be as slow during cooking as it was in the thawing process.

— Parts of the bird may be cooked tender without having reached temperatures high enough to destroy all organisms and spores. Especially will this be true if the fowl was grossly contaminated to begin with, since cooking temperatures cannot be expected to denature all toxins that may be present.

— Any added ingredient such as dressing in and around the bird will compound the insulation problem.

As a precaution against cooking problems, it is a good idea to have the oven thermostat checked at regular intervals to ensure that desired temperatures are actually reached. Remember also to allow adequate oven space to permit a normal cooking rate. Overcrowding may slow heat transfer appreciably. Similarly, garnishes, dressing, and other accoutrements in and around the turkey will serve as insulators.

Frozen products are not alone in requiring care in preparation. Any vulnerable food product which requires cleaning, trimming and portioning prior to preparation must not remain out of refrigeration longer than necessary, in order to keep microbic action to a minimum. To emphasize this point, let's consider typical problems which may arise in the preparation of a beef stew.

The Beleaguered Beef Stew

In addition to beef, the stew will contain various chopped vegetables, bases, and condiments, and the beef itself will likely have been trimmed from larger cuts. In approaching his many small tasks, the preparer will be tempted to say to himself, "I'll go ahead and leave this out of the refrigerator because I'm going to use it before long." The "before long" may become hours. The potential growth-time for organisms, added to the possibility of additional contamination at each step of preparation, becomes very significant. The vegetables, regardless of their source or the manner in which they have been handled, must also be thoroughly cleaned. Remember where *C. perfringens* organisms are found.

Prudence and good planning in making this stew can avoid an accumulation of hazards. If the meat portions are not to be used immediately after being cut up, they should be placed in covered, shallow containers and held under refrigeration. Vegetable components should also remain under refrigeration until the last minute. Perhaps the meat stock could be cooking, and the vegetables added to the stock as they are cleaned and cut.

In working with menu items such as stew, the cook often has to deal with large volumes, and this may introduce a problem. Because of the high water content of stew, temperatures above the boiling point of water, 212°F (100°C), can never be reached unless pressure cooking is employed. Thorough heat penetration can be a problem, too, especially in a large stock pot. Interior parts may never reach temperatures sufficient to destroy bacteria. It is important to recognize that heat transfer factors associated with insulation properties of a mixture may create a temperature condition in which bacteria can flourish. The best remedy for this is to stir frequently so that every particle of the mixture is completely cooked.

The "Battered" Breading

Chicken, steak and oysters to be breaded or similarly prepared may present a different problem. In this example, assume that the item has been cleaned, refrigerated, and otherwise readied for breading. At this point, food ingredients prepared separately are added, a batter made primarily of milk and eggs, we'll say. A cook on the alert for food contamination accidents would ask himself: when was this batter prepared, has it been under chill since that time, and how much use has it had? The same questions would apply to the crumbs, flour or meal the dipped item is to be coated with. The reasons behind these questions are simple enough:

— Milk and egg products provide a highly favorable medium for bacterial flora.

— The batter, even though it had a low microbic population when mixed, may have become contaminated through handling and contact with the dipped meat or seafood.

— The breading itself, while a relatively lean microbic medium in the dry state, may become a rich medium after exposure to drippings or carry-over from the batter.

— The breading operation is normally accomplished in a general work area, and under these circumstances the ingredients will be exposed to temperatures favorable for bacteria.

— Breadings and similar coatings are efficient heat insulators and may inhibit the cooking process.

All ingredients involved in this operation must be considered contaminated and potentially hazardous, and should therefore be refrigerated or discarded. Extreme caution must be exerted by the person preparing the food to minimize its exposure to unfavorable temperatures or contaminating conditions. If the food is not to be cooked immediately, return it promptly to the refrigerator.

Numerous other examples of procedures for pre-preparation and supplementary preparation could be offered, but the foregoing examples will provide some essential guidelines. A fundamental point to be remembered is that *conventional cooking procedures cannot destroy all bacteria and their spores or denature their toxins.*

There is more to be said about the effects of temperature in the preparation of uncooked foods and of food mixtures not to be cooked further after their components have been mixed. But before doing this let's consider some other points governed by the time-temperature principle: Refrigeration — keeping food at temperatures below 45°F (about 7°C)

— serves to retard or halt the proliferation of micro-organisms. There are two other effects of refrigeration which may not, however, be fully recognized: (1) Refrigeration suppresses the natural enzyme action in organic substances as well as the action of enzymes associated with micro-organic invaders or other contaminants. (2) Refrigeration reduces the rate of oxidation of food materials, animal or vegetable. Neither of these phenomena is likely to result in a hazard to health, but either one may cause a marked deterioration in the culinary quality of food.

THE TEMPERATURE HAZARD IN UNCOOKED FOODS

Both of the above considerations — safeness and culinary excellence — come dramatically to the fore in the preparation of salads. Salad vegetables are not likely to support extensive microbial flowering, but they may harbor large numbers of more or less dormant organisms. The handling of salads becomes critical when other ingredients which do support growth — like dressings, sauces and ground meat — are added.

For example, suppose we make a sandwich spread from the turkey previously cooked. The celery and other raw ingredients, let's say, have been taken from the refrigerator, cleaned thoroughly and chopped. The meat, already boned and finely cut, is stirred into the mixture. The mayonnaise or salad dressing is then added. Any of these ingredients, or the act of adding them, may introduce contaminants in our sandwich spread. In any event the mixture has now become a rich substrate for bacteria.

Salad dressings often have egg and vegetable oil as their principal constituents, as well as vinegar or lemon juice and seasonings. The vinegar of lemon juice adds acetic acid which is *not* favorable for bacterial growth, but once the mayonnaise or salad dressing is mixed with the other ingredients, the acid content of the salad may be diluted to a level which no longer retards bacteria. In many cases the additional nutrients will, in fact, accelerate microbial action and oxidation.

A similar condition will result in sandwiches spread with mayonnaise or salad dressing. In this case the bread, meat and other ingredients serve to reduce the acidity. Sandwiches and salads often become dangerous because they are prepared in large quantities and held for prolonged periods at bacteria-incubating temperatures. Even when the sandwiches are refrigerated the bread acts as a heat insulator.

SAFE HOLDING CONDITIONS

After the foodservice manager has prepared his principal menu items, he has to give serious attention to holding and storage. "Hot" foods must be held at 140°F (60°C) or higher. If the stew of our example is removed from the range and placed immediately on a heated steam table, there will probably be no major problem, provided adequate equipment is available and is used properly. But the manager should be alert to the following requirements:

— Ensure that all parts of the food are exposed to a safe holding temperature. If a deep vessel is used, the lower parts may be sufficiently heated, but be aware that upper surfaces are being cooled by the surrounding air.

— Containers must be covered or otherwise protected against splash and spillage.

— Proper utensils must be used for portioning or serving, as follows: Cups, bowls and other utensils without a handle should be avoided. Use a long-handle ladle, dipper, or the like, that keeps the server's hand away from the food. The serving utensil should be stowed in a proper place when not in use. If left in the pot, its handle may contaminate the food. If set down carelessly it may become cross-contaminated.

In some operations, soups, stews, salads and such are portioned by the table server or by the customer. These situations increase the risk of contamination unless strict attention is given to the workability of this procedure and its use is carefully monitored. It should become "instinctive" for servers to make a visual check of serving dishes and tableware for apparent soil or improper cleaning. Supervisors should spot-check from time to time to encourage safe practices by server and patron.

It is important that the food contact surface of dinner plates not be touched or handled in any way in portioning or serving. With hard-to-hold ware such as bowls, this may be difficult but proper layout of the serving area and storage of dinnerware can be of considerable help. Plates, bowls, tumblers, cups and silverware can be transported and stored in such a fashion that the handler need not touch food-contact surfaces at any time.

Our problem turkey will provide an example of a rather special handling problem. In order to debone and slice the turkey after cooking, it was necessary to cool the bird down. Once the meat was sliced, it was held in the refrigerator until serving time, which chilled it even more. In holding and serving the turkey, it is important to note:

— That food should not be prepared further in advance than necessary. Even under the best of conditions, extended refrigeration or hot-holding of foods will not improve their culinary quality, and excessive handling always increases the risk of promoting bacterial action. This is especially true of food that is changed from hot to cold, and then reheated.

— That most hot-holding equipment is intended for just that — to receive foods already heated to the desired temperature of 140°F or higher, and to hold them at that temperature. The equipment is usually not designed to receive cold or cool food and heat it to temperatures lethal for micro-organisms.

Holding breaded, fried, and baked dishes presents some different problems. Holding temperatures of 140°F (60°C) or above will tend to dry out or overcook the food. One is therefore inclined to hold these items at lower than approved temperatures. If it is necessary to keep food below 140°F (60°C) to preserve its flavor, the food should be prepared as needed and served right away. When handling dry foods, the temptation is to take the shortcut and pick them up with the fingers. Food servers and customers must be discouraged from this practice. Appropriate tongs or other utensils should be provided. This rule applies particularly to rolls, bread, butter pats, and ice.

Another aspect of serving requires constant vigilance: the inclination on the part of servers, busboys and others to handle clean place settings and serve food without washing their hands after wiping tables and bussing soiled dishes. The potential for contamination in these practices is extremely high, and workers must be made keenly aware of the hazards. Planning and scheduling of tasks can eliminate a large part of this problem, and supervisors must see to it that such conduct does not go unchallenged.

After a meal has been served, the food-service manager may think that his worries are over. If the menu has been sold out, this may be largely true. If not, there are still some decisions to be made and some work to be done. Mainly it's a matter of deciding what to do with "leftovers." The storage and handling of leftovers can lead to all kinds of pitfalls. Keep these considerations in mind:

— Regardless of the care exerted during preparation and serving, leftover foods have undoubtedly been exposed to contamination.

— The cumulative effect of holding foods in the temperature danger zone in the process of heating and cooling presents a formidible hazard.

— Some highly vulnerable menu items like custards and puddings should not be held over at all.

Take, for example, the beef stew prepared earlier. Suppose that 16 gallons of stew had been prepared, and that four gallons were served, leaving 12 gallons to be held over until tomorrow. Too much time, effort and expense have been tied up in this stew to discard it, so the decision is made to keep it. It's time to recall that the cooking process does not destroy all organisms, and that the product was exposed to further contamination during serving. It is essential that the stew be held above 140°F (60°C), or be chilled and held below 45°F (about 7°C) as soon as possible. Since it is seldom practical to hold the stew above 140°F (60°C) for an extended period, the most practical solution is to hold it in the refrigerator. And, because commercial refrigeration equipment is not intended to receive large quantities of hot food, the stew must be cooled beforehand.

With these factors in mind, some arithmetic on the cooling and refrigeration of this stew is in order. Twelve gallons of stew will occupy approximately 1-1/2 cubic feet of space (2770 cu. in.) and weigh about 93 pounds. If this stew is placed in a stock pot with a diameter of 16 inches, it will be about 13 inches deep. The center of the stew will be about 6½ inches from the top or bottom and 8 inches from the sides of the container. If this pot of stew were placed in a refrigerator which has an air temperature of 40°F (about 4°C), it would require well over 36 hours to cool the center portion to below 50°F (10°C). *That's too long!* There would be more than enough time for dangerous bacterial growth, so it is obvious that a better cooling procedure must be used.

Water is a much better heat conductor than air. One way to achieve proper cooling would be to divide the stew into two 6-gallon lots and put them both in an ice-water bath to chill. To chill stew uniformly, it must be stirred frequently. The pots of stew must be covered as soon as practical and protected from addi-

tional contamination. By using the ice-water bath, with frequent agitation, the temperature of the stew can be brought down from 140°F (60°C) to around 75°F (about 24°C) in an hour or so. The refrigerator can accommodate it much better now, but the "mass" or "depth" of the food is still to be considered. We must ensure that the stew is further cooled, below 45°F, as soon as possible. Further stirring under refrigeration will aid in achieving the desired condition.

The cooling time for a given mass of food can be calculated with some precision, but for our present purposes it is sufficient to emphasize one very important concept, and we will use the example of the 16-inch stock pot to illustrate it. If there were two inches of stew in the pot, the center would be located one inch from top or bottom and eight inches from the side. Let's assume it would require two hours to bring the temperature at this center from 75°F to 50°F (about 24°-10°C). Using the same pot, the same refrigerator and the same stew at 75°F, assume that it is filled to a depth of four inches. Perhaps your immediate reaction is that since it is twice as deep to the center it will take twice as long to cool it to 50°F, or four hours. This is *not* the case. The cooling rate in relation to depth is proportionate to the square of the depth of the food mass, or in this case four times as long — 8 hours. If the pot is filled to a depth of six inches, it will take nine times as long to cool, or 18 hours; eight inches deep, 16 times as long to cool, or 32 hours, and so on. See Figure 15. What counts then in calculating cooling time is the relative depth to the center of the food. This is the reason items to be

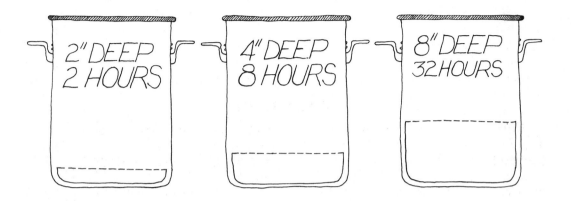

Fig. 15 Using the same pot, the deeper the food mass the longer the cooling time. But the change in cooling rate is not linear. In algebraic terms the cooling rate **CR** varies as the square of the thickness **T** (in this case the depth), as follows:

$$CR = \tfrac{1}{2}(T^2)$$

chilled should be placed in *shallow* containers to expedite the cooling process.

In the case of our stew, it could be poured into two 18″ x 24″ pans to a depth of about three inches for relatively rapid cooling.

Large quantities of soup, gravy or other highly-fluid liquid foods can be cooled quickly in a steam-jacketed kettle by introducing cold water, instead of steam, into the jacket. The cooling can be further accelerated with a mechanical agitator.

IN SUMMARY

Important considerations for the protection of foods in preparation, holding and service:

— Positive action to prevent contamination that may lead to foodborne illness is a fundamental requirement in foodservice sanitation.

— Time-and-temperature control is the most important single factor in controlling food contamination once it has occurred. Perishable and potentially hazardous foods should not be held at room temperature longer than is absolutely necessary.

— Form a mental image of a food thermometer and, between the readings 45° and 140°F (7°-60°C), color it red! This is the danger zone in which harmful bacteria are most likely to multiply.

— Almost all frozen foods can safely be cooked in the frozen state and this is usually the preferable cooking method. The exceptions are food products which require further cleaning or preparation.

— Cooking does not destroy all bacteria and spores, or neutralize all toxins secreted by bacteria.

— Raw foods must be processed separately from cooked and ready-to-eat foods to avoid cross-contamination. Cutting boards, knives, slicers, grinders and utensils used for processing raw foods must be sanitized before being used for cooked and ready-to-serve menu items.

—Each time we add a new ingredient to a food mixture we introduce another possible source of contamination.

— Safe foodhandling reaches a climax when we are ready to serve the food. One of the most serious hazards lies in holding hot foods at moderate serving temperatures.

Gastronomic considerations can never take precedence over the dictates of safety. Any conflict between sanitary and culinary quality must always be resolved in favor of the customer's health.

Unpredictable customer demands will eventually frustrate the most forehanded meal planning. When his timing is off, the foodservice manager may be caught in the dilemma between over-cooking a menu item or having to re-refrigerate it. The costly retreat to the refrigerator will often be his best way out.

Human hands are the principal carriers of foodborne disease agents. Direct handling of food and food-contact surfaces should therefore be reduced to a minimum.

The geometry of food and the physical principles involved in heat transfer play a major role in safe foodhandling operations. Let these factors work for you in heating and chilling food rapidly and uniformly to minimize dwell-time in the temperature danger zone.

A CASE IN POINT

Tim swore at the weatherman as he contemplated the big kettle of split-pea soup. It was supposed to be a cold, blustery day — split-pea weather — and he had planned the menu accordingly. As it turned out, Citrus Fruit Sunburst would have been more appropriate.

"One man's meat is another man's. . . ." He cut it off with a shudder and returned to the steaming pot of soup. It was half way through the dinner period and almost no takers. A super soup it was, too — made from a rich stock sporting tasty chunks of ham, and seasoned to a tee — with any luck, a real winner. He pondered his options:

1. Hold it at 160°F and pray for a late horde of split-pea lovers.
2. Let it cool and heat each serving as needed.
3. Let it cool and refrigerate it for tomorrow.
4. Discard it.

The soup was made early that afternoon, held at a simmer for nearly an hour, and was now standing at about 170°. Further holding at this temperature would soon dry it out and wreck its flavor.

What would you do?

Tim held it at 160° and prayed. Of the alternatives he gave himself that was the only one acceptable. No. 2 would work for a while, but time and mild temperatures would be against him, allowing bacteria to flourish. No. 3 would also be risky because of excessive cooling time for a large food mass. No. 4, a costly surrender. Assuming the soup was worth it, Tim did have another alternative: Quick-chill in small batches and refrigerate it now; reheat for any split-pea lovers that might straggle in. Tomorrow would be another day.

MORE ON THE SUBJECT

For further reading the student is referred to the following sources described in the Bibliography, Appendix B.

Reference   4   *Quantity Food Sanitation* (Longrée, 1968), Chaps. VIII-XIII.

5   *Sanitary Techniques in Food Service* (Longrée/Blaker, 1971), Part III.

12   *48 Ways to Foil Food Infections* (Bete, 1970).

15   *Hot Tips on Food Protection* (HEW).

19   *Discover the Unseen World — Prevent Food Poisoning* (MSU, 1966).

23   *Food Service Sanitation Manual* (PHS, 1962), Part V.

24   *Current Concepts in Food Protection* (USDA, 1973).

25   *Protecting Our Food* (USDA, 1966). "Meals Away From Home."

26   *Emerging Foodborne Diseases* (Bryan, 1972).

27   *Foodborne Diseases of Contemporary Importance* (Bryan, 1973).

33   *Food Service Operations* (USN, 1971), Chap. 2.

PART THREE
THE SAFE FOOD
ENVIRONMENT

CHAPTER 7

Sanitary Facilities and Equipment

Maintaining sanitation standards in an improperly designed foodservice is like trying to carry water in a colander. You waste a lot of time and energy without accomplishing much. A relatively simple job becomes hard and frustrating, and the natural human reaction is to settle for something less than the desired result. In this chapter we will discard the colander and consider an approach with fewer holes in it.

A facility designed, built and equipped for cleanability is more likely to be kept sanitary. The fewer the places for soil, pests, and micro-organisms to collect, the easier it will be to keep the establishment clean. And the easier it is to maintain cleanliness, the closer we will be to achieving a contamination-free environment. Any spot that cannot be readily cleaned will very soon become infested.

Building cleanability into a foodservice plant is a process that should begin on the drawing board — whether the project involves modernization, expansion, or new construction. It is at this early stage, of course, that we can most easily plan for ideal sanitation conditions. But what about the manager who inherits an established foodservice? He can seek cleanability every time he buys a new piece of equipment, or repairs or remodels his facility. In any event the goal is to provide a significant degree of "built-in" sanitation in every structural and mechanical feature.

We will examine four major aspects of the foodservice plant into which sanitation capability can and should be integrated:

- **Materials for floors, walls and ceilings**

- **Design features which facilitate cleaning and maintenance**

- **Design features which facilitate pest control**

- **Design of utilities to prevent contamination**

Materials and conditions that promote cleanability sometimes run afoul other design criteria. Sanitation requirements and a desirable decor detail may, for example, not be completely in accord. Heavy drapes lend a rich feeling to a dining room, but they are dust collectors and are hard to clean in place. In designing the facility, management must, above all, safeguard the health of the public. At the same time, a setting must be provided which is appealing and conducive to pleasant dining.

Cleanability and the Law

In most communities, the sanitary design of a foodservice is largely governed by law. And, as always, ignorance of the law is no excuse.

Depending on the type of construction, governmental controls may involve public health, building, and zoning departments. Some jurisdictions provide for review and approval of plans prior to new construction or extensive remodeling. Local health authorities often provide check lists of sanitation features they deem necessary or desirable. While the architect or contractor usually obtains the required permits and approvals, ultimate responsibility rests with the owner-operator of the facility. He will have to make sure that the plans and specifications are complete. To do this he may need to familiarize himself with design standards, and, in some degree, even the language of the building trades.

MATERIALS

Materials for floors, walls and ceilings should be selected for ease in cleaning and maintenance, as well as for appearance. But too often they are chosen on the basis of how they look when first installed. That is a short-sighted attitude, of course, if the good looks quickly vanish under the first layer of stubborn grime. In judging any covering material, the manager should ask himself: Can it stand up to anticipated wear and tear? Is it so porous or absorbent that it will retain soil and resist ordinary cleaning? Does it have a smooth surface, or is it full of crevices that will hold dirt? Can it be painted or otherwise finished, if desired? Will it be serviceable over a long period, or will it require frequent refinishing, touching up or replacement?

Flooring

The choice of floor coverings — and the price range — is almost unlimited: wood; carpeting; asphalt, rubber, cork, vinyl asbestos and vinyl tile; quarry and ceramic tile. Each has its special advantages and disadvantages; each is appropriate for use in some places and inappropriate in others. A number of considerations should govern the choice of a specific floor covering:

— Comfort, quietness underfoot, cleanability, safety (non-skid) and ease of maintenance or replacement.

— Resistance to wear and marring, grease, cleaning compounds, and color changes.

Each area of a foodservice imposes its own requirements for floor coverings, and common sense will often indicate which belongs where. Ceramic tile is out of place in a dining room, and carpeting is just as illogical a choice for a dry-food storage area. In the same way, different floor materials are needed for food preparation areas, restrooms, the scullery and the inside garbage storage room.

In most operations, the kitchen imposes the greatest demands on floor coverings. Quarry tile with 15% abrasive (for traction) and waterproof joints is generally recommended. Terrazo decks, in contrast, may be too slippery. Frequent water and grease spills in food preparation areas obviously make carpeting and wood floors unsuitable, and are also destructive to most resilient coverings. Water seepage between the joints can "lift" some resilient tiles. Others do not wear well when exposed to grease, oil, solvents or strong soaps.

Except in food preparation and storage areas, carpeting has found favor because it absorbs sound and shock. Although it is available in fabrics easy to clean and inexpensive to maintain, carpeting does require daily vacuuming and regular shampooing to remove soil.

Walls and Ceilings

Some of the same factors in judging floor coverings also apply in the selection of wall and ceiling materials. Comfort and safety are of less concern, but ease of cleaning, noise reduction, and maintenance are important.

Ceramic tile is a popular wall covering for application in almost every area of the foodservice. The grouting should be smooth, waterproof and continuous — with no holes to collect soil or to harbor vermin. Stainless steel is often used in food preparation areas high in humidity and subject to considerable wear and tear.

Painted plaster or cinder block walls are appropriate for relatively dry areas if they are sealed with soil-resistant and easy-to-wash glossy paints — epoxy, acrylic enamel and similar materials. Toxic paints must never be used in food preparation or storage areas, and plaster should be avoided in high moisture environments.

Ceiling coverings which improve light conditions and absorb sound are, of course, the best selections. Plastic or coated panels of other materials are good substitutes where plaster and conventional finishes may not be practical or desirable. Often the panels can be cleaned in place or removed for cleaning.

Dining rooms, bars and lobbies ought to have an appealing appearance to patrons, but interiors chosen only for good looks may backfire. Drapes, sconces, pediments and fixtures that soil easily or are hard to reach can become eyesores as they gather dust and cobwebs. The manager who does not specify the need for cleanability and accessibility in his instructions to the interior decorator will very shortly regret that omission.

DESIGN FOR CLEANING AND MAINTENANCE

In the ideally designed foodservice establishment, every place is reachable with a scrub brush, broom, mop or vacuum cleaner. Where a cleaning tool cannot reach, or where it doesn't reach easily enough, soil and vermin are sure to collect. Inaccessible areas can result from bad planning and construction and from deterioration of building materials —e.g., cracking plaster or a settling floor. Generally competent design and construction criteria will normally serve the purposes of sanitation, but a foodservice facility requires special emphasis on certain features, as follows:

— No sharp corners or edges. (Rounded conformations are preferred, to minimize dust accumulation. Rounded coving is indicated where floors and walls meet, to eliminate tedious scrubbing and probing of crevices which could harbor insects and harmful bacteria.)

— Location of equipment and furnishings with a view to avoiding narrow spaces. (Adjacent walls should be far enough apart to allow entry of cleaning tools. Adequate space must be provided under equipment and furnishings to facilitate cleaning. Some equipment may be sealed to the floor, wall-mounted, or rigged on wheels.)

— Strategic placement of utility outlets for use with power cleaning equipment. (For example, if quarry tile with abrasive is the floor covering in the kitchen, conventional wet mopping and squeegeeing by hand will not be desirable. Instead, vacuuming after manual or mechanical brushing will be necessary, calling for convenient electrical outlets. Handy utility hook-ups are also required for pressure spray equipment.)

Equipment and Fixtures

The design of equipment is, of course, no less important than its placement in facilitating sanitation. It is ironic that a piece of sanitation equipment would itself be so designed as to contribute to unsanitary conditions. But take a dishwashing machine with elaborate contours and complicated protuberances, for example. Crevices, raised surfaces, and inaccessible areas may harbor soil, insects, and micro-organisms, which could be transferred to sanitized tableware.

How do we select equipment on the basis of its cleanability? The following list of preferred characteristics is based on recommendations of the National Sanitation Foundation:

— Equipment is easily disassembled and cleaned. All parts coming into contact with food products are readily accessible for examination and cleaning, or can be easily removed.

— Materials which come in contact with food are nontoxic and impart no significant color, odor or taste to food; are non-absorbent and inert to food products and cleaning compounds.

— Corners and edges are rounded off. Metal surfaces meet with smooth, easily cleaned seams.

— Food contact surfaces are smooth and free of pits, crevices, ledges, inside threads and shoulders, bolts and rivet heads.

— Gaskets, packing and sealing materials are installed so as to be easily cleaned; are not affected by food products or cleaning compounds, are non-toxic and non-absorbent.

— Splash zones are designed and constructed for cleanability. There are no indentions or embossing to hold soil, insects or micro-organisms.

— Coating materials — especially those used on food contact surfaces — resist cracking, chipping and spalling.

— A permanently attached plate describes cleaning procedures.

— Waste liquids and condensation can be removed without difficulty.

One piece of equipment deserves specific mention because it is so commonly used and can so easily become a source of contamination. We refer to the cutting board, a kitchen item as traditional as the chef's hat. Repeatedly chopped and scored with sharp implements, the cutting board inevitably becomes crisscrossed with hideaways for contaminants. Hard rubber blocks have found acceptance in some instances. Wooden cutting boards should get a thorough cleaning regularly, and should be resurfaced or discarded after long use.

Design for Pest Control

Insects and rodents are relentless adversaries, forever trying to find a home in food environments. It is not so surprising that insects can pass through and inhabit extremely small spaces. But rodents can also make use of very small openings. A rat can squeeze through a hole only an inch wide. With this in mind, the manager must plan his establishment for pest control as well as clean-

ability. In Chapter 10, we will describe in some detail the hazards presented by pests, and the various methods for discouraging infestations. For now, it will be sufficient to discuss how we can prevent their entry.

Obviously, doors, windows and air intake ducts must be screened to exclude insects and rodents. Outside doors should be snug-fitting and self-closing. Less obvious, perhaps, is the need to check and tightly seal all cracks and openings in the building's foundations, and around heating, plumbing and electrical conduits. Anytime an opening is made in walls, floors, or ceilings to permit installation of such lines, a very real possibility exists that enough space has been left for entry of pests.

Leg-mounted equipment should be at least six inches off the floor so that floor cleaning equipment can be used under it. If the unit is large, it should be a few inches higher so that areas farthest to the rear can be reached.

To facilitate cleaning, some units can

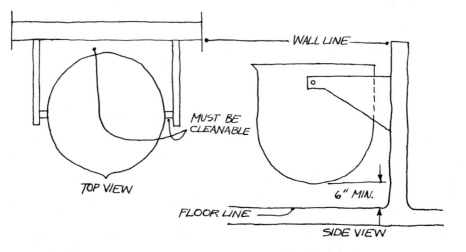

Fig. 16 Wall mounting allows free access for cleaning under a large kettle.

be made mobile by the use of wheels or casters. Others may be supported by a bracket attached to the wall (cantilever mounting, Fig. 16) in such a way that allows cleaning underneath. To wall-mount equipment designed to be free-standing may, however, invite other problems. For example, the unit may have a rounded splash-back top which is quite desirable when accessible for cleaning, but forms an uncleanable gully when attached to a wall.

Fixed equipment not mounted on legs also may harbor insects because of the spaces likely to result where the base meets the floor. Sealants should be used to provide an unbroken insect barrier, but there are some precautions to be observed. Only food grade sealants should be used in pantries and kitchens. And a sealant should not be used to cover up wide gaps caused by faulty construction. Such buried mistakes will ultimately be exposed, opening up new cracks to insects.

Design of Utilities

It seems that nothing enters or leaves a foodservice without having some impact, however indirect, on sanitation. Air, water, lighting, garbage and sewage disposal — all of these services require a lot of planning. Plumbing installations must provide for a safe as well as an adequate water supply. Sewage lines must be installed to preclude any possible contamination of foodstuffs. Lighting must be bright enough to clean by, and fixtures must be safe for use near foods. Ventilation must be sufficient to promote comfort and remove airborne micro-organisms. Garbage must, of course, be disposed of in a safe manner.

Plumbing and Sewage

In almost all American communities, plumbing design is closely regulated by law — and with good reason. In few areas of a foodservice can improper design pose such potential dangers. Since plumbing supplies us with water and also carries away wastes, very great care must be taken to see that the two never meet.

They can meet, however, in a system with a *cross-connection,* or in a system liable to *backflow.* A piping connection between the water supply line and a line carrying non-drinking water or other liquids, steam, gas or chemicals, is an obvious hazard. Backflow, in which contaminated liquids are allowed to flow back into the water supply, is possible in such common foodservice equipment as coffee urns and dishwashing machines because they have direct water supply connections. It can also occur if a water supply faucet is installed below the overflow line of a washer. Any time the wash basin is filled to the brim the faucet will be submerged (Fig. 17). *Back-siphonage,* a kind of backflow, occurs when there is a drop in water pressure. Should there also be an improper connection — say, to a garbage can washer — contaminated water can be siphoned back into the water supply.

If backflow plumbing defects cannot be corrected, backflow prevention devices should be installed. In any event, the manager needs to know enough about possible plumbing problems to recognize them and have them corrected. A periodic check should be made of all water outlets to ensure that defects have not been added to the sys-

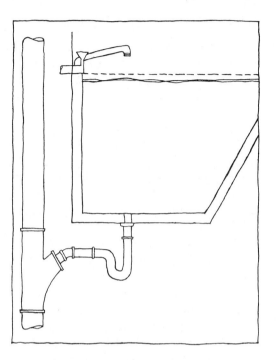

Fig. 17 Correct design to prevent backflow into water supply.

tem. A worker may, for example, connect a hose to a bib or faucet and leave the open end in a tub or vat. Backflow will develop unless the hose is promptly removed.

Sewage is the most dangerous reservoir of pathogens that can be found in the area of a foodservice, and the plumbing system must be foolproof against contamination from such a source. Overhead waste-water drain lines should always be avoided. A leaking sewage line in a food preparation or storage area could be disasterous. Sewer backups may also contaminate foodstuffs stored in an exposed basement for example. Any area subject to heavy water exposure — as a kitchen cleaned by hosing down — should have its own floor drain.

Ventilation

Sufficient ventilation is not only important for the comfort of workers and patrons; it is also a valuable tool for better sanitation. A properly designed ventilation system can help remove airborne bacteria, odors, smoke, grease vapors and moisture — without creating drafts. Excessive condensate on kitchen walls and ceilings will promote the growth of molds and bacteria. Contamination can obviously occur if droplets fall into unprotected foods.

A new and increasingly important aspect of foodservice ventilation concerns what happens to the air after it has been exhausted — in short, air pollution. Exhaust air from cooking areas may be thick with food odors, smoke and grease, and have to be purified. Clean air ordinances are on the increase. Their provisions range from controlling odors to eliminating smoke. The wise manager will check his installation against local regulations to make certain he is in compliance.

Lighting

Lighting must be sufficient for employees to do their jobs effectively — and that includes cleaning the establishment. It is human nature to overlook dirt that cannot be easily seen. The location of lighting fixtures is also important in minimizing contamination that may result from shattered glass bulbs or fluorescent tubes. A fixture located directly above food being prepared or held on the serving line should be shielded with metal globes or sleeves.

Water Supply

An in-town foodservice operation usually has no major problem in obtaining a safe water supply. But no matter where the facility is located, there may be variations in water pressure. When the pressure drops, it can play hob with the efficiency of an automatic dishwasher so that tableware will not be cleaned as well as it should. The alert manager will keep tabs on the adequacy of the water supply and take extra precautions when the pressure is reduced.

A private water supply, such as an individual well, should be regularly inspected to determine that it is safe. Most local health departments will perform this analysis on request.

Providing enough hot water of proper temperature seems to be a universal problem, not only with foodservice operations, but with most establishments serving the public. Too often, new demands are placed on the hot water supply — through remodeling and expansions — without regard to heater capacity.

Equally important in the consideration of a water heater is its recovery rate (the speed with which the heater produces hot water), the size of the holding tank and the location. To maintain the right temperature for sanitizing — 180°F (82°C) — it might be necessary to use a booster unit near the dishwashing machine. If the water heater is located a long distance from equipment using the hot water, the addition of a recirculation pump should be considered. The pump will not only speed the flow of hot water, but it will also cut costs by reducing the amount of water and heating energy used.

Garbage and Trash Disposal

From the sanitation point of view, it is essential that garbage and trash be removed from food preparation, dish washing and food storage areas as soon as possible.

If a mechanical food-waste disposal unit is used in the dishwashing system, food wastes from the serving area will be disposed of during the scraping and washing action. Before installing such a disposal system, a check should be made with the local authorities to determine whether its use is authorized — whether sewer lines are adequate to handle the wastes.

Garbage and wet trash should be removed either to refrigerated garbage storage rooms or to docks or platforms provided outside the building for temporary holding pending scavenger pickup or other disposal. Garbage should be held in plastic bags and in covered containers both in kitchen collection areas and at temporary storage points.

"Dumpster" units containing garbage and trash should be kept closed, and litter should not be permitted to remain in the vicinity of containers, incinerators, and garbage and trash storage points.

There are two systems which reduce the volume of trash, and both are finding increasing use. *Pulpers* grind refuse into small parts which are flushed with water. The water is then removed so that the processed solid wastes can be trucked away. *Mechanical compacting* of dry bulky wastes, such as cans and cartons, is particularly valuable in establishments that are cramped for space. This process reduces garbage volume to as little as one-fifth. Use of a compactor requires access to a drain, water for cleaning, and usually a power source.

Incineration may be provided for disposal of burnable trash and sometimes garbage. Incinerators must meet local air pollution control ordinances with respect to emission and odor control. Usually, the cost of providing incinerators capable of burning garbage within air quality control standards makes it prohibitive to dispose of garbage in this manner. Incinerators should not be used as a temporary collection container for wastes to be burned.

Any garbage processing or disposal area will attract vermin. Strict housekeeping and sanitation practices are necessary to minimize bacterial action and prevent insect and rodent infestation.

Toilet Accommodations

Local building and health codes usually specify how many lavatories, water closets and urinals are required by a foodservice to meet the need of diners and employees. Toilet and handwashing facilities for employees may be located in their dressing and locker rooms.

Soap dispensing and hand-drying equipment should be provided in sufficient number and in locations convenient to lavatories and washstands. Remotely operated liquid soap dispensers are strongly recommended for employee washrooms. Bar soap has a limited acceptability. Either cloth or disposable paper towels are acceptable, but the cloth toweling should be of the automatic, self-retracting type. Warm air dryers are considered the most sanitary, but some workers find them too slow working and may resort to the forbidden practice of wiping their hands on their aprons.

IN SUMMARY

The foodservice which is hard to clean will in all probability not be cleaned well. Sanitation efforts will be greatly facilitated if the establishment is designed and equipped with cleanability in mind.

The manager should give cleanability first priority when he plans a new foodservice, or when he remodels or expands an existing one. In most communities, plans for new construction or extensive remodeling are subject to review and approval by local regulatory agencies.

Four construction features lend themselves to "built-in" sanitation in a foodservice facility: (1) Materials used in the floors, walls and ceilings. (2) Equipment and fixtures. (3) Design features for pest control. (4) Design of utilities.

Avoid the inclination to select floor, wall and ceiling coverings solely on the basis of good appearance. In the long run it will pay off to consider their cleanability and durability. Much the same can be said of equipment to be used almost anywhere in a foodservice operation. Crevices or surfaces that catch dirt can also harbor micro-organisms and vermin. Equipment should be placed so that all areas around and under it are accessible to cleaning apparatus.

The foodservice with built-in sanitation will also have its utilities designed for cleanability and freedom from contamination hazards. Included in this category are plumbing and sewage installations, ventilation and lighting systems, water supply and drainage systems, water heating equipment, garbage disposal systems and toilet facilities.

A CASE IN POINT

"Hey, Dad, what's salmonella — some kind of ice cream?" Tim's inquisitive 10-year-old son was perched on a stool in the restaurant kitchen.

"It's nothing to eat," Tim said, wishing the boy would change the subject.

"Well, what does it have to do with food?"

Tim sighed. With equipment sanitation in mind, he had started a survey of the most likely contamination traps in his kitchen, but at the moment he was in a trap himself. One way or other the boy would have the whole story, so he began:

"Salmonella is a tiny germ that gets into food, and people who eat the food can get very sick."

"How do you keep it from happening to our customers?" the boy challenged.

Tim thought for a moment. "Well, we make sure everyone washes his hands after he goes to the toilet. And we see to it that everyone uses clean utensils in working with food.

"Most germs are killed when food is cooked," he explained, "but sometimes germs get into cooked food. Then you're in trouble unless there are very few germs and you can keep them from growing. That means not letting food stand at mild temperatures."

He opened the door of the refrigerator. "We also make sure no one puts egg custard beneath a cut-up raw chicken. The water from the chicken could drip into the custard and the custard doesn't get cooked any more.

"For another thing, we keep our cutting boards wire-brushed and clean so germs can't grow in the slash marks."

"I guess you have a few things to worry about, don't you, Dad?"

Tim ran his hand through his son's shaggy hair. "And we don't let people work around here unless they keep their heads covered," he said, slapping a chef's hat on the boy's head.

"Gee, I might just grow up and run a restaurant like you," the boy said.

"If you do, remember to watch out for salmonella."

Reaction?

In an informal way Tim covered the field very well, including direct contamination, cross-contamination and progressive contamination under dangerous holding temperatures. We could add a sensational cross-contamination case in point involving a hard-to-clean meat slicer: Contaminated meat served in a Sioux City, Iowa, restaurant in a two-week period in 1970 resulted in 250 reported cases of salmonellosis traced largely to a meat slicer sequestering numerous colonies of *Salmonella enteritidis* in inaccessible parts of the machine. (Ref. 39)

MORE ON THE SUBJECT

For further reading the student is referred to the following sources described in the Bibliography, Appendix B.

Reference   7   *A Self-Inspection Program for Foodservice Operators* (NRA, 1973).

10 "The Sub-Standard Washroom" (NRA, 1966).

20 *Manual on Sanitation Aspects of Food Service Equipment* (NSF, 1968).

21 *Standards on Food Service Equipment* (NSF), Standards 1 - 8, 12, 13, 18, 20, 25, 26, 33, 35, 36, 37, and C-2.

22 *Sanitary Design and Evaluation of Food Service Equipment* (Farish, 1971).

23 *Food Service Sanitation Manual* (USPHS, 1962), Part V, Sections D and E.

28 *Environmental Health and Safety in Health-Care Facilities* (Bond/Michaelsen/DeRoss, 1973), Chap. 11.

CHAPTER 8

Cleaning and Sanitizing

We know at this point why it is important to have a clean, sanitary foodservice. Every method for combatting dangerous micro-organisms and pests, for keeping food safe and wholesome, is founded on the rules of good housekeeping. Still to be discussed is the subject of how we do that vital job—what we need in the way of tools, materials and procedures.

We have already explained the concept of "clean versus sanitary," but the distinction bears repeating here because of its importance in the actual cleaning and sanitizing operation. An object which is "clean " is free of visible soil but may be contaminated with hazardous bacteria. A "sanitary" object may look dirty but does not carry harmful micro-organisms. As we will find, it is difficult to make an object sanitary if it is not first cleaned. Although the two may seem to be the same, the techniques and materials used to achieve clean and sanitary conditions are quite different.

A clean *and* sanitary establishment is the result of a planned program, properly supervised and followed on schedule. Unfortunately, when the going gets rough and workers are rushed trying to meet the needs of customers, correct practices are often neglected. Only a manager who is knowledgeable and alert can prevent these breakdowns in good sanitation discipline. He must know how a cleaning/sanitizing job should be done, instruct his workers accordingly and then see to it that his instructions are carried out.

Aside from the continuous cleanup tasks in kitchen and serving areas, there are two general approaches in organizing the maintenance of a foodservice establishment:

— "Clean your own" or "Clean as you go" and . . .

— "Scheduled" cleaning at the hands of a specialist.

The first is largely self-explanatory. Each employee is responsible for his own work area or station and for the equipment he uses. "Scheduled" cleaning by a specialist takes care of everything else —floors, walls, carpets, loading docks, dishwashing equipment, refrigerators, storage rooms, etc.

How you apply these methods in your establishment will depend on its size and the type of service provided. A combination of the two will often be employed.

Cleaning compounds and equipment of many kinds are available. We will describe the general types in some detail, and discuss recommended techniques and procedures in using them. The subject matter will cover these major aspects:

- **Selection of detergents for specific cleaning jobs**

- **Basic types of cleaning equipment**

- **Sanitizing procedures**

- **General kitchen cleaning procedures, by hand and machine**

DETERGENTS

A detergent is broadly defined as any cleaning agent — including sand, steam and water — which removes soil. More specifically, it is a manufactured chemical substance that makes cleaning easier. Water, the universal solvent, is of course a good cleaning agent if applied with sufficient force. Since enough pressure is often not available or practical, a detergent is added to the water

to make up the deficiency.

Detergents generally in use in the foodservice industry are chemical compounds or mixtures which are usually prepared for specific applications — on floors, in dishwashers, for removing mineral deposits, etc. A particular detergent should be selected for its special cleaning properties. A compound which is powerful in one application may prove totally ineffective in another.

For greatest effectiveness, these four requirements must be fulfilled in cleaning with a detergent:

— The detergent must be brought into close contact with the soil to be removed.

— The detergent must loosen and remove the soil, then keep it suspended in the cleaning solution.

— Soil must be rinsed away completely to prevent its being redeposited on the cleaned object.

In addition to being an effective cleaning agent, a detergent must also be compatible with its intended use and with the facility. It must be an efficient water conditioner, noncorrosive, non-toxic, or safe when used as directed, and economical in cost. Since some detergents are more effective than others, the quantities required to do a job should be considered in making a cost comparison.

Detergents are classified according to their composition and intended use. *General purpose* detergents are mildly alkaline and effective for most soil removal and suspension. *Heavy duty* detergents, such as those needed for machine dishwashing, are highly alkaline. They break down protein soils and fats. *Acid* detergents are used where alkaline compounds fail — in attacking mineral buildup from water and milk, for instance.

Thus far, we have been dealing with chemicals for cleaning purposes, but there are detergents which also contain sanitizing agents. They are especially useful in treating heavily soiled objects or areas, such as waste receptacles and loading docks, where food contamination is not a problem. Observe manufacturer's recommendations scrupulously if these products are to be used in food-contact areas.

Finally, there are even more specialized agents for oven cleaning, tarnish or stain removing, scouring and polishing. Many of these compounds are toxic to humans and, in common with other cleaning agents, should always be stored away from food preparation areas.

Consult your suppliers to determine which chemical compounds are needed for your specific applications. Don't depend on the advice of a foodservice operator down the street. Detergents which work well in his place may not be suitable in yours. Each establishment is different, with individual soil loads, menus, degrees of water hardness, physical layouts, tableware and equipment.

CAUTION Under no circumstances should you attempt to formulate your own detergents. Unless you know exactly what you are doing — unless you have the knowledge of a trained chemist — the result could be very dangerous to your employees and patrons, and damaging to your facility. At best, you will probably produce a compound which does not achieve the result desired. *Always read the labels on any product used for cleaning or sanitizing and closely follow the manufacturer's instructions.*

GENERAL TYPES OF DETERGENTS

Highly alkaline	Heavy duty cleaning. Removal of wax, baked-on grease in ranges or ovens. Used with dishwashing machines and power sprays.
Mildly alkaline	General-purpose cleaning for floors, walls, ceilings, equipment and utensils. Manual cleaning.
Acid	Removal of mineral deposits, water "stone," milk "stone" and hardwater films. De-liming dishwashing machines and similar equipment.
Detergent/ sanitizers	Highly contaminated areas and objects — loading docks, garbage cans. Non-food contact surfaces. Food contact surfaces, *at recommended concentrations only.*

CLEANING EQUIPMENT

Providing adequate and appropriate sanitation tools is the responsibility of management. There are plenty of exotic labor-saving devices on the market, but whether they will be useful and economical in your facility is a decision you must make. Differences of opinion about cleaning equipment and tools often arise between management and workers. The final responsibility in purchasing belongs to the manager, of course, but he should take into account the recommendations of the man who will use the machine. Again, the special needs of the establishment may be the governing factor.

Use of wire brushes, steel wool and coarse abrasives — anything which will mar the surface being cleaned — should be avoided. Once a surface is scratched, another harborage for soil is provided which may be difficult to clean with ordinary detergents.

Brushes of natural fibers, nylon or plastic are probably the most common cleaning tools, and they are used in manual as well as mechanical operations. The type selected is a matter of individual preference, but it should be noted that some wood-backed fiber brushes are themselves hard to keep clean. Nylon and plastic brushes with composition backing, in contrast, are easily cleaned and ideal for a good many hand or mechanical cleaning operations. They have strong, flexible, uniform bristles which wear well, are non-abrasive and do not absorb water.

Other cleaning tools on your shopping list will include cellulose sponges, metal scrapers, heavy rubber gloves, clean cloths, brooms (vertical and push-type), wet mops, dust mops, double buckets with mop presses, and dust pans.

For the heavier cleaning jobs, there are more sophisticated devices: vacuum cleaners (wet and dry), wet vacuums with spray attachments, and automatic scrubber-vacuums. The needs of your establishment must determine whether it is practical and economical to invest in these machines. If you do, you will also have to train workers in their operation and maintenance, and provide intelligent supervision, to make the expense worth while. All too often, costly machines are not used to full effect because of poor maintenance, misapplication and misunderstanding of operating procedures on the part of employees.

CLEANING PROCEDURES

To guide workers, cleaning and maintenance procedures must be prepared — in writing — regardless of the size of the facility. If this is not done, probably cleaning will not be adequate and equipment will be abused or used improperly. The instructions ought to be brief and written in basic, understandable language so that workers are encouraged to follow them. Written procedures should include what is to be cleaned, when or how often it is to be cleaned, how the job should be done and what materials and equipment should be used. Suppliers of cleaning agents often furnish model procedure forms, and the operator simply "fills in the blanks" with specific instructions. The following is an example of such a procedures form:

KETTLES, STEAM JACKETED
DAILY

1. Flush the kettle with warm water immediately after use, and allow to drain.

2. Close valve. Fill kettle to $\frac{1}{4}$ full with hot water. Add   oz. of (cleaning agent). Brush-wash all surfaces inside and out. Use probe and brush to clean draw-off pipes and outlet valves as the solution is draining. Scrub adjacent piping, braces and valves.

3. Rinse all washed surfaces with clear, hot water.

AS NEEDED

If any scale or film remains on the kettle after the daily cleanup, it must be given a special de-scaling treatment. Proceed as follows:

1. Fill kettle with warm water to just above the normal liquid level; turn on the steam.

2. While water is heating, add   oz. of (cleaning agent).

3. Bring solution to near boil. Brush-wash all surfaces above the liquid level as well as outside surfaces wherever scale has developed. Allow the hot solution to stand in the kettle for one hour, brush above the liquid level occasionally during the soak period.

4. Open drain valve and brush off all loosened scale and film as the kettle empties.

5. Rinse all surfaces with hot water.

SANITIZATION

The term "sanitized," you will recall, means that the bacterial contamination of an object or surface has been reduced to a safe level. It is a step above "clean" — which is merely the absence of soil — but a step below "sterile" — which is the absence of all living organisms. Sanitizing materials are referred to as germicides, bactericides, disinfectants, etc.

Sanitization is accomplished with heat or chemical processing, but it will not be effective unless the articles or surfaces to be sanitized are first physically clean. Caked-on soils not removed by cleaning, for example, may shield bacteria from a sanitizing solution. Sanitization, in short, is no substitute for good cleaning.

A conscientious manager's reaction to all this might be: "Why don't we just sterilize everything and be on the safe side?" The practical answer is that the cost would be prohibitive.

Heat Sanitizing

Exposing a clean object to sufficiently high heat for a sufficiently long time will sanitize it. Generally, the higher the heat, the shorter the time required to kill harmful organisms.

Pasteurization of milk is one example of sanitization. It involves the application of heat, with the key elements of time and temperature in correct proportion. Originally this was accomplished with a temperature of 143°F (about 62°C) for 30 minutes. But imagine the chaos in the average foodservice kitchen if all utensils to be sanitized were tied up for half an hour! The alternative, of course, is to increase the heat and reduce the time. *Immersion of dishes or utensils in water at 170°F (about 82°C) for not less than one-half minute* is now generally acceptable to most regulatory authorities. Since these time and temperature requirements may vary for specific situations and conditions, it is well to check the regulation in force in your locality.

Still another means of heat sanitization is through the use of live steam or flowing hot water. It is important to note, however, that the temperature at the surface of the object being sanitized is what counts, not the temperature of the water or steam in the line. Water loses heat very quickly, especially when sprayed on a cold surface.

Chemical Sanitizing

Chemical compounds with sanitizing as well as detergent and deodorizing properties have found a wide acceptance for routine cleaning purposes in a foodservice. The three most commonly used compounds employ chlorine, iodine or quaternary ammonium (quats).

The chlorine and iodine compounds have much in common: They will kill most bacteria if used correctly, are not greatly affected by water hardness and lose effectiveness in water that is too alkaline. Because of this last factor, it is necessary to apply a thorough rinse to items that have been cleaned with general-purpose detergents, which are alkaline, before applying the sanitizers. There are also some differences between the two classes of compounds: Iodine compounds (iodophors) have a built-in indicator of concentration — the stronger the solution, the deeper its amber color. Chlorine compounds are more likely than iodophors to be harmful to the skin, and to attack metals.

Quats are relatively non-corrosive and are effective in both acid and alkaline solutions. The bacteria-killing power of a particular quaternary compound may be limited, but a blend of quat compounds can dispatch a greater variety of micro-organisms. A high degree of water hardness (over 200 parts per million) will make quats less effective. However, they do have better detergent and deodorizing properties than chlorine or iodine compounds.

Some sanitizing agents are toxic and are therefore acceptable for use only on non-food surfaces. Others may not be toxic, but they impart undesirable flavors and odors and are unfit for food-service use. To make certain you have not selected sanitizers in this category, carefully study the manufacturer's descriptions and directions before using the compounds.

Concentration, Time and Temperature

The following table (8 - 1) provides general guidelines for sanitization by im-

Table 8-1

Agent	Concentration ppm or mg/1	Time	Temperature
IMMERSION SANITIZATION			
Chlorine	50-100	1 min.	75°F+
Iodine (pH 5.5 or below)	12.5	1 min.	75-120°F
Quats	200	1 min.	75°F+
CLEAN-IN-PLACE SANITIZATION			
Chlorine	100-200	2-3 min.	75°F+
Iodine (pH 5.5 or below)	12.5-25	2-3 min.	75-120°F
Quats	200-300	2-3 min.	75°F+
POWER-SPRAY SANITIZATION			
Chlorine	200-300	2-3 min.	75°F+
Iodine (pH 5.5 or below)	25	2-3 min.	75-120°F
Quats	200-300	2-3 min.	75°F+

mersion, for cleaning in place (CIP) and for spray-type applications. When in doubt, check with your local health authority.

Sanitization compounds are shipped in stock concentrations. The manufacturer's instructions will usually include directions on how to dilute them properly. Since the proportion of active components in sanitizer formulations is usually very low, less than one per cent, it is expressed in ppm (parts per million) or mg/1 (milligrams per liter).

As the table indicates, concentrations necessary for a complete kill are progressively higher for CIP and Power-Spray than for the Immersion method. Temperature ranges are also important in effecting sanitization. The solutions must be warm enough — at least 75°F (about 24°C) — to allow chemical reactions to take place. But if temperatures are over 120°F (about 49°C), chlorine and iodine may leave the solution.

CLEANING IN THE KITCHEN

Regardless of whether a manual or mechanical cleaning job is involved, it will not be fully effective without adequate preparation. Any kitchen cleanup task should start only after these three steps have been taken:

— By preflushing or scraping, remove as much gross physical soil as possible before using detergent. Simply dumping dirty utensils into a tank of cleaning solution is a waste of detergent and water. Excessive soil "ties up" the detergent so it cannot work as intended.

— Utensils with baked-on or hardened soils may have to be soaked before scrap-

ing. Don't rush the job by using a wire brush.

— Use clean washing solution.

— Use the right detergent. There is a difference between "sink" and "machine" detergents, and they are not always interchangeable. Sink solutions are designed to allow soils to fall to the bottom. Machine compounds suspend the soils in solution.

Manual Cleaning

Water and "elbow grease" were the original cleaning agents. Without sheer physical effort, the kitchen of yesterday's tavern or inn just didn't get cleaned. There is still a lot of manual effort involved in cleaning, but modern tools and materials make the work easier. A three-step procedure is recommended for most cleaning and sanitizing:

1. Wash with a clean detergent solution at 100-120°F (about 38-49°C). This will be effective, and not too hot for the worker's hands.

2. Rinse in clear, warm water.

3. Sanitize by one of the methods already discussed.

You may contemplate varying the basic wash-rinse-sanitize process, but it would be best to consult your local health official before proceeding. Not only will he tell you if your proposed procedure is safe and acceptable, he may also have some valuable suggestions to offer on detergents and sanitizers.

Since many implements undergo heat sanitization in kitchen use, it may seem unnecessary to treat them further. "Why sanitize a pot if we're only going to cook in it?" is a question operators often ask. Two more good questions answer it:

What assurance is there that the pot will be used immediately for cooking? How do you know that the cooking time and temperature will be enough to overcome contamination of the entire pot?

Don't take chances. Clean and sanitize every utensil, food contact surface and unit of equipment after each use. That way there will never be any reason for concern about safeness.

Too often work tables and stationary equipment only get a quick swipe with a damp cloth. They, too, require regular cleaning and sanitizing to minimize bacterial contamination and discourage insect infestation.

Machine Cleaning

In some applications, manual cleaning is irreplaceable, but it does have its limitations. Properly operated and maintained, machine washing is more reliable in removing soil and bacteria from tableware and kitchen implements. Because of this factor and the needs of high volume operations, the foodservice industry has increasingly moved to the use of dishwashing machines. Even so, purchase of washing equipment can represent a sizeable investment, and your individual requirements should be carefully considered before making a decision. You

will need:

— A machine which has adequate capacity to handle your work load of soiled table- and kitchenware.

— A water heater of enough capacity, properly installed as close to the machine as possible.

— An efficient layout in the dishwashing area to utilize manpower and machine to best advantage.

— Workers who know how to operate and maintain the equipment, and are knowledgeable about the correct use of required detergents and other chemicals.

— Regular inspection by management to make certain that correct procedures are being followed.

— A covered or enclosed storage system that will protect dishes from contamination after they have been cleaned.

An adequately-sized dishwashing machine which complies with National Sanitation Foundation Standard No. 3 and is properly installed and maintained will perform satisfactorily. NSF Standard No. 3 also contains a detailed guide for detecting and solving dishwashing machine problems. Since clean dishes are of paramount importance to the foodservice manager, the guide is presented here in its entirety. (Table 8-2)

Table 8-2	DISHWASHING PROBLEMS AND CURES	
Symptom	Possible Cause	Suggested Cure
Soiled Dishes	Insufficient detergent.	Use enough detergent in wash water to ensure complete soil removal and suspension.
	Wash water temperature too low.	Keep water temperature within recommended ranges to dissolve food residues and to fa-

Table 8-2   DISHWASHING PROBLEMS AND CURES (cont'd)

Symptom	Possible Cause	Suggested Cure
		cilitate heat accumulation (for sanitization).
	Inadequate wash and rinse times.	Allow sufficient time for wash and rinse operations to be effective. (Time should be automatically controlled by timer or by conveyor speed).
	Improperly cleaned equipment.	Unclog rinse and wash nozzles to maintain proper pressure-spray pattern and flow conditions. Overflow must be open. Keep wash water as clean as possible by pre-scraping dishes, etc. *Change water in tanks at proper intervals.*
	Racking.	Check to make sure racking or placement is done according to size and type. Silverware should always be presoaked, placed in silver holders without sorting. Avoid masking or shielding.
Films	Water hardness.	Use an external softening process. Use proper detergent to provide internal conditioning. Check temperature of wash and rinse water. Water maintained above recommended temperature ranges may precipitate film.
	Detergent carryover.	Maintain adequate pressure and volume of rinse water, or worn wash jets or improper angle of wash spray might cause wash solution to splash over into final rinse spray.
	Improperly cleaned or rinsed equipment.	Prevent scale buildup in equipment by adopting frequent and adequate cleaning practices. Maintain adequate

Table 8-2   DISHWASHING PROBLEMS AND CURES (cont'd)

Symptom	Possible Cause	Suggested Cure
		pressure and volume of water.
Greasy Films	Low pH. Insufficient detergent. Low water temperature. Improperly cleaned equipment.	Maintain adequate alkalinity to saponify greases; check detergent, water temperature. Unclog all wash and rinse nozzles to provide proper spray action. Clogged rinse nozzles may also interfere with wash tank overflow. Change water in tanks at proper intervals.
Streaking	Alkalinity in the water. High dissolved solids in water.	Use an external treatment method to reduce alkalinity. Within reason (up to 300-400 ppm), selection of proper rinse additive will eliminate streaking. Above this range external treatment is required to reduce solids.
	Improperly cleaned or rinsed equipment.	Maintain adequate pressure and volume of rinse water. Alkaline cleaners used for washing must be thoroughly rinsed from dishes.
Spotting	Rinse water hardness.	Provide external or internal softening. Use additional rinse additive.
	Rinse water temperature too high or too low.	Check rinse water temperature. Dishes may be flash drying, or water may be drying on dishes rather than draining off.
	Inadequate time between rinsing and storage.	Allow sufficient time for air drying.
Foaming	Detergent. Dissolved or suspended solids in water.	Change to a low sudsing product. Use an appropriate treatment method to reduce the solid content of the water.
	Food soil.	Adequately remove gross soil before washing. The decom-

Table 8-2   DISHWASHING PROBLEMS AND CURES   (cont'd)

Symptom	Possible Cause	Suggested Cure
		position of carbohydrates, proteins or fats may cause foaming during the wash cycle. Change water in tanks at proper intervals.
Coffee, tea, metal staining	Improper detergent.	Food dye or metal stains, particularly where plastic dishware is used, normally requires a chlorinated machine washing detergent for proper destaining.
	Improperly cleaned equipment.	Keep all wash sprays and rinse nozzles open. Keep equipment free from deposits of films or materials which could cause foam build-up in future wash cycles.

Types of Dishwashing Machines

Dish- and ware-washing machines of many types and capacities are available. Table 8-3 may help in selecting the right machine for your requirements.

IN SUMMARY

It is obviously important to keep a foodservice establishment clean and sanitary, but how is the manager to do that? He can require each worker to maintain his own area, or he can employ a cleaning specialist, but in most cases he will use a combination of these two approaches.

To make sure that correct procedures are followed, the manager should regularly inspect cleaning activities, adequately train workers and provide easy-to-understand written instructions.

A variety of cleaning products and equipment items are on the market and management must know which ones are suitable for the individual needs of the establishment. Use of the wrong material or equipment for a particular application can prove unnecessarily expensive and even harmful.

Detergents are available for a number of soil removal purposes, and there are also sanitizing procedures and chemicals for reducing bacterial populations to safe levels.

Some materials are toxic to humans. For safety and effectiveness, each should be used in accordance with the manufacturer's recommendations and label instructions.

A three-step process—wash, rinse, sanitize—is necessary to make items clean and sanitary. Use of a sanitizing agent will not be fully effective if it is not preceded by cleaning and the rinsing away of detergents.

Table 8-3

Ware-Washing Machines

Specifications for temperature, water volumes, times, and pressures

Type of Machine	Wash			Pumped Rinse			Final Rinse @ 20 PSI (Flow Pressure)		
	Vol. Water	Min. Exposure	Min. Temp.	Vol. Water	Min. Exposure	Min. Temp.	Minimum Water Volume	Min. Exposure	Min. Temp.
Single Tank Stationary Rack									
16 x 16 inch	60 gal.	40 sec.	150°F				1.15 gal.	10 sec.	180°F
18 x 18 inch	75 gal.	40 sec.	150°F				1.44 gal.	10 sec.	180°F
20 x 20 inch	92 gal.	40 sec.	150°F				1.73 gal.	10 sec.	180°F
Single Tank Stationary Rack Single Temp.	60 gal.	40 sec.	165°F				14.7 gal.	30 sec.	165°F
	75 gal.	40 sec.	165°F				18.6 gal.	30 sec.	165°F
	92 gal.	40 sec.	165°F				23 gal.	30 sec.	165°F
Single Tank Stationary Rack Chemical Sanitizing	Total 80 gallons includes sanitizing rinse		120°F	(Not applicable except in multiple-tank machines, which recycle the rinse water. *Know your machine* before attempting to adjust tank temperature. "Multiple-tank", refers to machines with wash and pumped-rinse cycles — not to systems with mechanical pre-wash or pre-scrape units. The latter do not contribute to sanitization, due to low water temperature.)			Total 80 gallons includes wash vol.		120°F, 50 ppm Cl$_2$ or other accepted sanitizing solution
Single Tank* Conveyor 20-inch width	3 gal./lin. inch conveyor	15 sec.	160°F			160°F	6.94 gal. per min.	Max. conv. speed 7'/min.	180°F 6" spread 5" above conveyor
Multiple Tank* Conveyor 20-inch width	1.65 gal./lin. inch conveyor	7 sec.	150°F	1.65 gal./lin. inch	7 sec.	160°F	4.62 gal. per min.	15'/min. max. convey. speed	180°F 3" spread 5" above conveyor

*Note: Wash and/or pumped-rinse makeup water may add up to 2 additional gallons/min. to final rinse volume demands.

A CASE IN POINT

Tim has a secret. And a sore spot when it comes to dish-washing machines.

His dishwasher unit went out, and for a while it looked as if they would have to close up shop. Fortunately, or call it what you will, it happened when Aunt Clotilda was in town for a visit.

Tim had walked away from the pandemonium in the kitchen and saw her as she turned in the front door. Aunt Tilly often flashed a bright smile, but a special twinkle came into her eye as Tim told her the news.

She marched to the back, looked around the stalled machine and immediately took command. She seized the operating manual from a worker, glanced briefly at the first page and set it aside.

"You city folk depend so much on gadgets," she said, "you can't do anything yourself. I hope you've got enough hot water." She had the busboy clear out a pot-and-pan sink and started in herself scraping and stacking a pile of breakfast dishes.

"Put these in to soak," she ordered. "Get some soap powder and scrub them with a good brush. I don't know about 'sanitize' but scrape, wash and rinse in scalding hot water has always done the trick where I come from."

The job was soon finished and Aunt Tilly was back out front, about to take her leave.

"See you later," she said to her nephew. "I've got some shopping to do. You'll be all right. But you'd better have that boy sweep behind the washing machine. I see a lot of dust and cobwebs back there."

Tim thought he noticed a wink — half a wink, maybe. He told her goodbye and went back to the kitchen. When no one was looking he walked over to the dishwasher and plugged the power cord in with a wry grin. He wished there were more Aunt Clotildas in this world.

Comment?

No comment.

MORE ON THE SUBJECT

For further reading the student is referred to the following sources described in the Bibliography, Appendix B.

Reference   3   *Food Poisoning and Food Hygiene* (Hobbs, 1938) , Chap. 11.

           7   *A Self-Inspection Program for Foodservice Operators* (NRA, 1973) .

      23   *Food Service Sanitation Manual* (USPHS, 1962) , Part V, Sections D and E.

      24   *Current Concepts in Food Protection* (USPHS/FDA, 1973) .

      28   *Environmental Health and Safety in Health-Care Facilities* (Bond/Michaelsen/DeRoss, 1973) , Unit II, Chap. 3.

CHAPTER 9

The Cleaning Schedule

It is a fact of life that foodservice facilities require constant cleanup. Every time a meal is served, the cleaning and sanitizing work that has gone before is at least partly undone. Each customer, directly or indirectly, leaves behind him a trail of dishes and disorder which must be quickly cleared away. The cleaning work is, quite literally, never done.

In the foregoing chapter we discussed materials and methods of cleaning and sanitizing. Here we are concerned with how often we clean and when we do it — the daily, weekly and monthly cleanup and maintenance routines for all equipment and spaces.

How often do you clean? The response might be: "Well, as often as necessary." A more realistic examination will reveal the need for a planned approach. If cleaning is done on demand, so to speak, it may never be done thoroughly. The natural tendency to put off cleaning until more pressing tasks can be completed will allow soils to collect until they are harder to remove and the work requires a disproportionate amount of time. A simple scrubbing with detergent will not remove the caked-on grease from a griddle back plate. The aged grime in a neglected corner of the kitchen will not yield to a quick swabbing down. The cleaning job will become too great, and the manager will find himself struggling to keep soiled areas and equipment at a barely tolerable level. Inevitably the day will come when it is absolutely necessary to clean, but workers may then be busy with other tasks. And that is when the danger of food contamination grows ominously large.

The intelligent alternative is scheduled cleaning. If peace of mind is not sufficient cause for scheduled cleaning, there are 10 other good reasons:

1. It provides a sound basis for efficient management and effective use of time. (It makes the manager plan ahead, so he is less likely to find himself short-handed in the midst of a crisis.)

2. It helps to distribute the work load fairly. (Employees will have less reason to think they are being forced to do more than their share.)

3. It ensures that necessary work is not overlooked ("Cleaning on demand," in

contrast, often ignores dirt that can't be readily seen.)

4. It avoids duplication of work.

5. It makes specific staff cleaning assignments. (Since there is no question about who is to do a job, it is more likely to be done conscientiously.)

6. It provides for substitute or relief workers. (Scheduled cleaning anticipates worker vacation periods and those tasks which require outside specialists.)

7. It forms a logical basis for supervision. (Since the manager knows exactly who is to do what, and when, he is in a better position to make his inspections.)

8. It sets goals for cleaning crews. (There is no lost time in debating what work is to be done next, or in organizing the job.)

9. It provides a useful training ground for new personnel. (They are taught the cleaning routines of a foodservice and, at the same time, come to know management's organized approach.)

10. It is helpful in developing a good public image. (It tells others—both workers and patrons—that orderly and dependable procedures are followed in the establishment.)

Scheduling cleaning is a much more involved matter than merely specifying when the work is to be done. As we will recommend in this chapter, planned cleaning requires:

— A written cleaning schedule which assigns responsibility for the work, indicates the time it is to be performed, and describes cleansing procedures and techniques for each space and unit of equipment.

— Management initiative in training workers and in conducting self-initiated inspections to make sure that proper procedures are being followed.

HOW TO DESIGN A CLEANING SCHEDULE

The scheduling of cleaning operations should be so comprehensive that no area or unit of equipment is neglected. One way to organize such a detailed schedule is to go from room to room in your establishment with clipboard in hand. Note all the areas—floors, walls, ceiling, shelves—which need cleaning on a regular basis. Now, for each room, identify all the units of equipment—coffee urn, walk-in reefer, work tables — that require maintenance. Based on your practical experience, try to estimate how often each of the items on your written list becomes soiled or contaminated enough to require cleaning. Also try to determine how many man-hours are required to do the work.

The schedule you draw up on the basis of your list will probably need to be varied to fit changes in menu, personnel and equipment. Cleaning schedules should be flexible enough so they can be adapted to all the business changes which afflict foodservices. Still, the standard plan will invariably need to specify these fundamentals:

- **What is to be cleaned**

- **Who is to clean it**

- **How to clean it, and**

- **When to clean it**

This chapter will not attempt to design a schedule for all cleaning and maintenance routines. There are too many

Table 9-1

Sample Cleaning Schedule

Food Preparation and Holding

	Unscheduled	Daily	Weekly	Seasonally
FLOORS	Wipe up spills immediately. (1)	Sweep and mop between meals. Pressure-clean mats nightly. (2)	Pressure-clean or brush-scrub. (2)	Reseal as required and inspect for needed repairs. (2) (3) (4)
WALLS and CEILINGS	Wipe off splashes immediately. (1)	Wipe areas around grill, range, prep sink between meals. (2)	Clean as indicated. (2)	Wash and scour (pressure-clean, brush or wipe). (2) Inspect for needed repairs. (4)
WORK TABLES	Wipe away debris and soils. Clean and sanitize between changes in use. (1)	Scrub and sanitize between meals. (2)		Inspect for needed repairs. (4)
HOODS	Empty grease traps as required. (2)	Clean inside and out. (2)	Remove filters and pressure-clean. (2)	Inspect and clean ducts. (2) Lubricate fan and motor if required. (3)
COOKING EQUIPMENT				
Broiler	Empty drip pan as required. Wipe down. (1)	Scrub shields, grids and outside apron. (2)	Clean thoroughly with degreaser. (2)	Inspect and service. (3) (4)
Fryer	Wipe down as required. (1)	Drain and filter fat. Clean vat and apron. (2)	Clean thoroughly with degreaser. (2)	Inspect and service unit. Check thermostat and timers. (3) (4)
Griddle	Scrape after each use. Empty drip pan as required.	Clean with polish stone. (1) Clean apron. (2)	Clean skirts with degreaser. (2)	Inspect and service. (3) (4)

Equipment				
oven	immediately. (1)	door and apron. (2)		(3) (4)
Oven	Wipe inside and out as required. (1)	Clean inside and out. (2)	Clean inside and out with degreaser. (2)	Service unit, check thermostats. (3) (4)
Range	Empty drip pans, wipe spills. (1)	Clean all surfaces, skirts and aprons. (2)	Remove burners and sections — degrease. (2)	Inspect and service. (3) (4)
Steamer	Wipe all spills immediately. (1)	Wash compartment and clean outside. (2)	Polish. (2)	De-lime if required. (2) Check gauges, thermostats and gaskets, and service unit. (3) (4)
HOLDING EQUIPMENT				
Refrigerator	Wipe all spills immediately. (1)	Check temperature. (4) Clean outside. (2)	Wash thoroughly. Remove and wash racks and shelves. (2)	Defrost and clean condenser coils. (2) Check temperature controls. (3) (4)
Steam table	Wipe all spills immediately. (1)	Wash inside and out. (2)	Polish. (2)	De-lime if required. (2) Check thermostats and heating elements. (3) (4)
FOOD PREPARATION EQUIPMENT				
Can opener	Wipe outside as required. (1)	Scrub and sanitize blade. (2)		
Mixer Peeler Slicer Grinder Chopper	Clean and sanitize after each use. (1) (2)	Clean and sanitize outside of units, stands, mounts, etc. (2)	Inspect for lubricant leaks and serviceability. (4)	Inspect, lubricate and service. (3) (4)

different types of foodservices, with too many different requirements and units of equipment, to allow for an all-encompassing schedule. We will, however, provide guidelines for scheduling cleaning and sanitizing operations which will be typical for many establishments.

Following the outline established above, your schedule should specify in easily understood instructions:

What—An individual schedule should be drawn up for each of the items listed in your clipboard survey.

How—Specific procedures were outlined in Chapter 8. The instructions should be written simply and in such detail that any employee—the veteran or the beginner—can easily follow them in doing a satisfactory job. Equipment manufacturers often supply sample cleaning instructions which can be adapted to your schedule.

Who—Is one of your own employees going to be responsible for the item, or will an outside cleaning specialist be employed? If the latter, who is responsible for requesting his services? Depending on the size of your operation, you may enter the name of the employee who is to do the work, or his job title.

When—Major cleanup is to be done at a time when contamination of foods is least likely to occur and interference with service is minimized. Do not schedule sweeping or mopping during the preparation of food, or, in the dining room, when patrons are present. Sweeping obviously raises dust, and mopping causes splashes, both of which can contaminate food being prepared or served to the customer. Just as obviously, a slick, wet floor poses a safety hazard during busy serving periods.

Special considerations—Perhaps safety or other considerations should be noted in your schedule. This includes posting barricades in areas where floors are "slippery when wet," or "do not use" signs when the hood filters are removed for cleaning. You will also want to enter cautionary notes for the dismantling and cleaning of potentially hazardous or easily damaged equipment.

Table 9-1 includes most of these factors in a sample cleaning schedule.

MANAGEMENT RESPONSIBILITIES

The manager has not fulfilled his obligations just by completing a cleaning schedule. Good sanitation is, among other things, a vigilant attitude on the part of a manager ready to provide training, supervision and follow-up. The manager is responsible for guidance and inspection to ensure that the job is performed according to schedule. But he must never regard his schedule as an immortal document which should be chiseled in granite. Instead, he should determine through continuing analysis of the operation whether refinements and modifications are needed. It is all very well to specify that floors be cleaned daily at 4:30 P.M., but what if the local garden club decides to have an extra-early dinner meeting in your place?

As was pointed out in Chapter 8, management has the responsibility to provide the proper tools and facilities for cleaning. The schedule in operation will surely tell whether management has met its obligations. In observing the functioning of a cleaning schedule, the manager should be prepared to answer for these situations:

— The schedule calls for wiping up spills and splashes immediately. Are the "tools" (clean wiping cloths and mops) readily available?

— The schedule calls for pressure cleaning. Are power and water supplies available for this purpose?

—The schedule requires separate cleaning and sanitizing steps. Is the equipment adequate to provide for both portable items and permanently installed units? If a large item that cannot be taken to a stationary vat is involved, are the appropriate pails and scrubbing tools on hand? Have the right chemical agents been supplied to workers?

To help develop the total cleaning plan, the manager may now combine the specific procedures described in Chapter 8 with his cleaning schedule.

Checking the Cleaners

Now that your cleaning schedule is functioning, you must still evaluate its effectiveness. Even the most conscientious and dedicated employees may depart from proper procedures in the rush of business or out of sheer forgetfulness. It is your job to see to it that deviation doesn't occur, and that means continual supervision and self-inspection.

To assist in this evaluation, use the National Restaurant Association's self-inspection system recapped in Appendix A. When and if self-inspection reveals defects in your cleaning schedule, don't be too quick to blame employees. It may be necessary to modify the schedule or make training more effective.

IN SUMMARY

A foodservice operation is forever in need of cleaning and sanitizing. These functions can best be performed according to an organized and scheduled program.

The cleaning schedule should be published in easily understood form and encompass all spaces and units of equipment.

The cleaning schedule should assign responsibility for the work, indicate how much time is required, note when and how often the cleaning is to be done and describe the procedures and techniques to be employed.

It is not possible to establish a cleaning schedule which will apply to all foodservices because requirements vary so widely.

As a general rule, cleaning should not be scheduled when it may result in the contamination of food under preparation or being served.

Management's responsibilities do not end when the cleaning schedule is established. The manager should provide continuing supervision, training and inspection to make certain that the schedule is being followed—and that it works.

A CASE IN POINT

It is clear from previous episodes that Tim is keen on cleanliness. It might even be fair to call him "Mr. Clean" himself. Tim has also shown himself to be an energetic organizer, but on this occasion things didn't pan out. The broiler didn't get cleaned.

Tim had a cleaning schedule, laid out according to the book. Or so he thought.

It told *What* was to be cleaned.

It told *Who* was to clean it.

It told *How* to clean it.

It told *When* to clean it.

Moreover, it was comprehensive. It covered every major and minor area and unit of equipment. It was written in clear-cut readable language that any employee could understand, and copies were delivered to every department head.

There was an adequate supply of tools and cleaning agents. No problem there.

And yet, the broiler didn't get cleaned. The job, mind you, was assigned to the cook's helper, a bright young fellow, newly hired, who in a brief time had shown himself a willing and able worker.

What happened?

The kitchen copy of the schedule didn't get posted and the cook forgot to tell his helper.

MORE ON THE SUBJECT

For further reading the student is referred to the following sources described in the Bibliography, Appendix B.

Reference 7 *A Self-Inspection Program for Foodservice Operators* (NRA, 1973).

10 *The Sub-Standard Washroom* (NRA, 1966).

13 "Attitudes About Sanitation in Restaurants," (*Food Service Digest,* 1966).

14 "Sanitation Checklists for Management," (*Cooking for Profit,* 1972, Part V).

23 *Food Service Sanitation Manual* (USPHS, 1962).

24 *Current Concepts in Food Protection* (USPHS/FDA, 1973).

36 *Programmed Cleaning and Environmental Sanitation for for Buildings, Plants, Offices and Institutions* (Soap and Detergent Association, 1971).

CHAPTER 10

Pest Control

Ever since flies were sent to plague the Pharoahs, we have been shooing, swatting and swearing at them. The rodent menace traces back to the earliest human records. But with today's advanced knowledge, there is really no reason to tolerate these and other pests at home or at work.

A foodservice manager can guard his establishment against insects, rodents and other vermin if he has an understanding of their feeding and breeding habits—and their vulnerability. We will discuss the most common of these pests, the hazards they present, and the most effective measures available to us in controlling them.

And control them we must. Most pests carry germs that cause disease. Insects and rodents also account for considerable profit losses through spoilage and waste, to say nothing of loss of patronage. A rat eats tremendous amounts of food for his size, but more important are the contamination and damage he causes. For example, he may nibble only one corner of a bun, but the entire package will probably have to be discarded because of the droppings and hairs he leaves behind.

Pests can be effectively controlled if we:

- **Prevent their entry by . . .**
  Eliminating small openings into the establishment.
  Disposing immediately of containers which may bring them in.

- **Withhold our "welcome" by . . .**
  Eliminating conditions which encourage nesting and breeding. (Cracks, crevices and improperly stored materials all can be used to advantage by pests.)
  Cutting off their food supplies. (Clean up "spills" and food scraps, keep garbage cans covered. In other words, starve them out.)

COCKROACHES

Cockroaches are one of the worst insect menaces which confront the foodservice manager. There are several varieties, all of them undesirable, ranging in size from a half inch to 1½ inches, and in color from pale yellowish-brown to black. While some subsist on vegetables, others consume almost any food.

Despite their difference, they have much in common. They are all persistent, hard to control, and difficult to eliminate. They carry a multitude of disease germs, and readily contaminate exposed foods. They have been responsible for spreading such serious diseases as typhoid fever and dysentery, as well as tapeworms. Their wastes cause foul odors.

Roaches are great hitchhikers. They or their eggs may be brought into the house in boxes, bags and cartons of produce, meat or other supplies. Or they may enter through small cracks in the building floor and walls. If a nearby tenant or building has roaches, then it's almost a certainty that the insects will try to infiltrate the foodservice.

They lay their eggs in dark, warm, well protected (hard to clean) areas. Narrow spaces in and between equipment and shelves are favorite harborages. They feed on dry or moist food, and a mere crust of bread can support a large roach population. They prefer the dark, but will come out into the light if they get hungry enough. They may be found virtually everywhere in a food establishment.

Preventing Cockroach Infestations

Good housekeeping practices are the best first defense against cockroaches. It's far easier to prevent them from getting started than it is to eliminate them once they've become established. Follow these four procedures to prevent cockroach infestation:

1. Remove all produce and meat crates and cartons as soon as possible, or keep them under refrigeration until you can.

2. Seal all crevices, cracks or other openings—especially where the wall meets the floor—for it is here that the cockroaches may enter, hide, or lay their eggs.

3. Clean cabinets, equipment, and shelves regularly. A favorite hiding place is under shelf liners or boxes left sitting on the floor.

4. Keep the premises well-lighted and ventilated to discourage these pests.

Contrary to a popular misconception, mice are not just baby rats, but are instead an entirely different species of rodent. From the foodservice manager's point of view, however, it's the similarities, not the differences, between the two that are important. Both rats and mice present similar hazards, and prefer similar living conditions. Some facts about these two scourges are in order:

1. Rats can get through an opening no larger than a quarter. Mice need an opening no larger than a nickel.

2. They both like the dark and areas where there is little noise or activity. They seek the same food and shelter conditions that humans like.

3. When they move about, they usually stay near walls and on the floor, and they repeatedly use the same pathways Knowledge of the location of these rodent runways is helpful in setting traps.

4. They build nests in dark, warm, quiet areas, and can cause extensive damage in building their nests.

5. They both have very sharp, strong teeth. Rats have gnawed through lead sewer pipes just to get a drink of water.

6. Rodents are notorious carriers of filth and disease organisms which they may pick up from such sources as garbage

and sewage. Often rats are literally covered with fleas which themselves can cause disease (plague) in humans.

7. One rat "pill" (dropping) may contain several million bacteria. Even if the pill doesn't get directly into food, it will dry, fall apart or be crushed. Then the particles may be blown or carried into food.

8. Given plenty of food and water, and a good place to build their nests, a pair of rats can have a litter of ten young about every other month. Mice can produce five or six young even more often.

9. Rats and mice carry food back to their nests, possibly from outdoor garbage to nests in the establishment.

Preventing Rodent Invasions

Rodents are secretive creatures, and they may be present in the establishment even if you don't actually see them. Look for the telltale signs: gnawing marks on food cartons, droppings the size of buckshot, body grease smears along their runways. Whether or not you suspect a rodent invasion, this prevention program should be followed:

1. *Keep rodents out.* Screen all outside openings. Make sure that doors and windows are in good repair and tight fitting. Fill in the spaces around pipes and wires where they pass through the walls. Close off any cracks or holes around the building foundation.

2. *Eliminate food, water sources.* Store food products away from walls and at least six inches off the floor. Cover and protect foodstuffs and potable liquids, promptly clean up food scraps and spills. Keep outdoor garbage cans on pipe racks or concrete platforms at least one foot off

the ground, and make sure that the cans are always tightly covered.

3. *Discourage nesting.* Promptly dispose of empty cartons and other waste materials which may harbor rodents. Regularly inspect for nests in any dark, warm, protected areas and in boxes where dry products are stored.

CAUTION. Never touch a dead or dying rat with your hands. If still alive, it may fight back and bite. If the rat is dead, then the fleas will abandon it as soon as the body temperature starts dropping.

FLIES

The fly of most concern in eating establishments is the common housefly. It thrives in the human environment. It flourishes on the same kinds of food and temperatures, and breeds and reproduces in the wastes left by people. To combat houseflies, it's useful to have an understanding of their habits and hazards:

1. They can enter a building that has openings not much larger than the head of a pin.

2. They rarely travel very far from where they hatch, but may be carried by air currents.

3. They have no teeth and must take their food in liquid form. (Watch one trying to eat sugar; it will spit on the sugar and let it dissolve, then suck it in. The spittle is swarming with bacteria, of course, and this is one of the most significant ways in which flies spread contamination.)

4. A female fly in a favorable environment can in one season produce offspring numbering in the thousands.

5. Moist, warm, decaying material protected from sunlight is required for the eggs to hatch and young maggots to grow. An uncovered garbage can is ideal.
6. They carry large numbers of bacteria on their legs and bodies, another major way in which they spread disease.

Controlling the Fly

It is obviously important to keep flies away from food, and that can be done most effectively by preventing their entry in the first place. Here are some ways to do that:
1. Use screens on all outside doors, windows and other openings, and keep all screens in good repair.
2. When receiving supplies, leave doors and screens open for the shortest time possible.
3. If an increase is noticed in the number of flies, check the garbage storage area as a probable breeding place. Is it always neat and clean, or is damp waste allowed to collect on the ground? Are the containers regularly cleaned outside and inside to eliminate hatching places? Are the containers kept covered, and are the covers tight fitting? (Even if flies hatch elsewhere, any accessible waste in your cans may attract them.)

OTHER INSECTS

Moths, weevils, and beetles are generally found in dry storage areas. You'll see the insects themselves, their webbing, clumped-together food particles, holes in food, and holes in packaging. There are other pests of lesser importance from a health standpoint, but they can still be expensive nuisances. In-

cluded are a variety of moths, ants, carpet beetles, silverfish, and firebrats. Again, good housekeeping and proper food storage are the best preventives for these pests.

BIRDS

Such birds as sparrows, starlings and pigeons may also present foodservice problems, mostly because of their droppings. In addition to being unsightly, the wastes may carry fungi which can infect humans. The birds themselves may be carriers of mites and a variety of diseases, including encephalitis and psittacosis.

Controlling Pest Birds

While it is not always a simple matter, any bird population can be effectively controlled. Good sanitation management will prevent the birds from being attracted by food on the property. Screening with small-mesh chicken wire, hardware cloth or non-staining nylon netting should be used to close windows, ventilators or doorways to nesting birds. If these methods fail, then more direct techniques are possible, within certain legal restrictions. Migratory birds are protected by federal law. Many state and local laws specify not only what birds may be controlled but also what methods may be used. The legality of any control methods should, therefore, be checked in advance with local conservation officials and representatives of the U.S. Fish and Wildlife Service.

Trapping is usually the most acceptable method of bird control. Pastes which make the birds uncomfortable, or wires that administer a mild electric shock are also effective in preventing birds from roosting. Both of these last two methods are fairly expensive and require frequent inspection and maintenance. Noisemaking devices and flashing lights have little effect on birds because they become readily accustomed to these irritants. If done repeatedly, such forms of harrassment as nest removal and the use of water sprays may be successful.

The most effective procedure is to employ a pest control operator who specializes in birds. He has the knowledge and equipment required for safe use of chemicals in combatting them.

Use of Poisons

When most people notice pests around the place, their first reaction is to call the exterminator. This is seldom a long-lasting solution to the problem because it doesn't eliminate the conditions which made the infestation possible. An important guideline reminds us that *Poison is no substitute for good sanitation.* This is good advice because:

— Any poison that will kill pests is probably also dangerous to people.

— Even if pests are eradicated with poisons, they will return unless sanitary maintenance is improved. The house must be cleared of food and environmental conditions that attracted the pests in the first place.

— Most pests develop a degree of immunity to pesticides.

Chemical Control of Insects

No foodservice operation will remain insect-free unless chemical control measures are continued on a regular basis.

It is an unfortunate fact of life in a dining facility that insects may return no matter how good the sanitation and housekeeping. But proper safeguards, aggressively applied, will reduce the severity of an infestation and make control measures more effective when re-infestation does occur.

Insecticides are available in concentrated or ready-to-use forms. They may be had in oil solutions, water emulsions, water suspensions, baits, and dry powders or dusts. Oil solution and water emulsion sprays are the most common types used in foodservice establishments.

OIL, WATER-BASED SPRAYS. Oil-based sprays are used where . . .

— Water might cause an electrical short circuit.

— Water might cause shrinkage of fabrics.

— Water may stain or spot wallpapers, etc.

— Water applications will not dry quickly enough to avoid mildew.

— The insecticide is oil soluble.

Water-based sprays are used where . . .

— An oil spray would create a fire hazard, as around hot ovens or pilot lights.

— Oil solutions would cause damage to such materials as rubber and asphalt tiles.

— The odor of oil is objectionable.

— The insecticide is soluble in water.

RESIDUAL, CONTACT, SPACE SPRAYS. Insecticidal sprays are designated as residual, contact, or space sprays, depending on the way they work and the method of application.

A *residual* spray is applied to a wall, floor or ceiling, and leaves a surface deposit which kills insects that contact it. The oil or water medium evaporates. To be effective, the treated surfaces must be made wet with a thin, uniform layer of the spray.

A *contact* spray must touch the insect to kill it. The spray is usually applied to a group of insects, such as a cluster of roaches in a corner or crack.

A *space* spray is a special-purpose contact spray. It is discharged into the air as a fog which drifts for several minutes. Space sprays are more effective against flying insects than crawling insects. Contact and space sprays usually kill insects more quickly than do residual sprays.

DUSTS. Insecticidal dusts generally contain in dry form the same toxic agents used in the various sprays. While dusts may be preferable to sprays in some situations, they require more skill and care in application and are best left to the professional pest control operator.

BAITS. These are combinations of an insect-attracting food, such as sugar, with an insecticide. Although not widely used in foodservices, they can be effective in controlling hard-to-reach ant and roach infestations, and in reducing outdoor fly populations. Since they are a poisoned food, however, special care must be exercised in their use and storage.

DO IT YOURSELF? The foodservice operator confronted by an insect invasion may be tempted to grab the handiest insecticide and go on the attack. Under some conditions that may indeed be the wisest, most economical course to take. If flying insects — mosquitoes, flies and gnats — are the problem, he can himself use an industrial aerosol "bomb." Or he can resort to a hand-operated space sprayer and one of the non-toxic insecticides, such as pyrethrum, or pyrethrum plus an activator.

In a number of situations, however, the manager may be far better off making use of a pest control operator. For example, the location and elimination of fly breeding areas may not prove as simple as it sounds, so professional assistance may be needed. Some infestations may be so stubborn that they just won't yield to an amateur's efforts.

Selecting the correct insecticide and using it properly is a more complicated procedure than might seem to be the case — calling for the help of an exterminator. Any insecticide carries a certain amount of peril — through accidental food contamination, for example — so it's wisest to use the least powerful chemical that can do the job. Some insecticides, or methods of application, have no place in a foodservice because they can't be used safely there. An insecticide that is effective against one type of insect may be ineffective against another, and picking the right one for the right bug can prove a bewildering experience for the untrained individual.

Finally, to complicate the selection problem, an increasing number of insects, in an increasing number of places, have acquired resistance or immunity to insecticides which were once effective. Flies and mosquitoes, for instance, have developed an immunity to DDT and related chemicals, and the German roach has become immune to chlordane.

A complete solution to the problem of growing immunity has not yet been found. But research is moving ahead. The National Pest Control Association, government agencies, and universities and other research organizations are investigating materials and techniques so that all types of establishments with pest problems may continue to operate safely.

Because this is a complicated problem, varying from pest to pest and with the section of the country, specific and unqualified recommendations on insecticides cannot be made. Where resistance to insecticides is found, the simplest procedure is to employ a professional exterminator.

WARRING ON RODENTS

On any list of nightmares which ruin the sleep of a foodservice manager, this one must rank highly: His establishment is crowded with diners. Suddenly, their conversations stop and their forks poise midway above their plates. A rat has been spotted scurrying through the room.

When that happens, it's an obvious indication that some cardinal rule of pest prevention has been violated. As with other pests, the best way to prevent rats and mice is to make the environment unsuitable for them. If available food and shelter are reduced, the rodent population will be proportionately reduced.

Keeping rodents out and starving them out, then, are the first lines of defense. But what do you do if these measures don't solve the problem? Supplemental measures include:

— Trapping, which is safe but sometimes slow. Spring traps are best, placed at right angles to rodent runways.

— Poisoning and gassing, which should be done only by the most experienced personnel for safe and efficient results.

— Ultrasonic devices which generate sound waves that rodents tend to avoid. The procedure is usually ineffective. When the rodents are hungry enough, they will even cross the sound barriers.

The foodservice manager shouldn't attempt any control measures beyond the placing of traps and using the safest of baits (those available to the general public). At the very least, use of incorrect poisons might kill a rat deep in its burrow, producing a stench that could linger in the establishment for months. A professional pest control operator, in contrast, has a choice of materials, methods and equipment, and he should know which to use where.

If the manager has to do his own exterminating, it's imperative that he exercise extreme caution in handling any poisons. Many authorities advise that red squill and anti-coagulants are the only poisons to be used by untrained persons, particularly in food-handling establishments. If these poisons are used, they should be obtained from reputable manufacturers who maintain good quality controls. Poisons must be stored away from foods. They must never be kept where there's a possibility they may accidentally get into foodstuffs or kitchen implements — say, above work tables, or in a dry food storage area.

PESTICIDES AND SAFETY

The safest assumption to make about pesticides is that anytime they are used or stored in the establishment, a hazard exists to the food supply, customers and employees. As already noted, pesticides may be toxic to humans, and some may also cause fire or explosion. Since the manager is ultimately responsible for the safety of his entire operation, he must always see to it that these precautions are being observed:

Precautions in Use

1. Be sure insecticide containers are properly identified and labeled.
2. Follow the advice of a professional exterminator. Have him perform those pest control functions which are hazardous and complex, and which he is better equipped to handle more safely, economically and effectively. Insist that any pest control operator you employ has insurance on his work to protect your establishment, employees and customers.
3. Read and follow package directions carefully. Use any pesticide only for the purposes intended.
4. Use the least toxic material that will do the job. And use it at the proper concentration level. When it comes to poisons, it's dangerous to reason that if one dose is good, a double dose will be better.
5. Avoid any possibility of contaminating food or food utensils with pesticides. Always consider the rule of nature that "if an accident can happen, it will."
6. Apply pesticides with care so as to minimize damage to wall coverings, floors, fixtures and other portions of the property.
7. Use the correct application equipment and be sure that it's free of residues from previously applied pesticides.
8. Wash hands after using pesticides to prevent transfer of toxic substances to the mouth. Do not spray aerosols near the eyes — propellent gas is very irritating.

Precautions in Storage

1. Never permit a pesticide to be transferred from its labeled package to any other kind of container. One of the most common causes of poisoning stems from the bad habit of storing pesticides in emptied food containers.
2. Store pesticides in a safe place — preferably in a locked cabinet — well removed from food handling and storage areas.
3. Aerosol "bombs" or other pressurized spray cans should never be allowed to become overheated — if they do, they may explode quite violently. Store in a cool place and don't expose to temperatures higher than 120°F (about 48°C).

Precautions in Disposal

1. Bags, cartons, bottles, or cans not under pressure should be disposed of as soon as they become empty. Containers for pesticides are a potential hazard even when "empty" because they probably still contain particles of toxic material.
2. Bottles and cans should be well rinsed in a disposal drain — never in a sink used for food, dishware or utensils — and then destroyed so they cannot be re-used. Paper and cardboard containers may be incinerated.
3. Break bottles and crush cans (not pressurized ones!), wrap in paper, and place in rubbish containers.
4. Because of the probability of explosion, empty aerosol cans should not be destroyed by burning. Any pressurized spray can should not be crushed in a compactor, or punctured.

5. Where a facility isn't available for destroying pesticide containers, they should be placed in covered trash containers which are not used for the disposal of edible garbage (swill).

IN SUMMARY

Insects and rodents have plagued man and destroyed his food supplies throughout recorded history. They continue to be a menace in the foodservice operation because they are carriers of foodborne illness and also reduce profits through spoilage and damage of foodstuffs.

To guard the foodservice establishment against pests:

(1) Physically prevent them from entering by closing off any openings in the building and by inspecting incoming supplies.

(2) Eliminate sources of food, water and breeding places through proper housekeeping.

Should insect or rodent pests become established, it may be necessary to resort to pesticides — but poisons should be considered a supplement, not a substitute for proper sanitation. While insect or rodent populations may be destroyed by pesticides, they will be sure to return unless the establishment is kept clean, and food and water supplies are protected.

Because most pesticides are toxic to humans, these poisons should be selected and applied with great care. To protect the customer, employees and the establishment itself, precautions must be observed in the use, storage and disposal of pesticides. While the foodservice manager may himself choose to apply pesticides, he should employ a pest control operator for the more complex or hazardous extermination tasks.

A CASE IN POINT

It was a bad day at Restaurant Rock.

Cockroaches!

Tim's doors were open to a wide clientele but these "guests" were strictly uninvited. He was beside himself. Only a week ago he had been reading up on pest control and had called away a general cleanup in every department: storerooms, kitchen, pantry and dining room. His people had vacuumed, scoured and checked every cubbyhole. Afterward he had called the crew together and made sure they knew the rules: Inspect all supplies — store promptly — clear out the empty crates — clean as you go — wipe up spills — don't let grease collect — keep food covered.

Tim even prided himself on the cleanest garbage cans in town.

And now this! Tim was stumped. Where could those creepy freeloaders have come from?

Would you believe from cockroach *eggs* in a produce carton? It is possible there was not a single cockroach in the house the day before. But the warm, dry crevices in a cardboard box are favorite nesting places for these pests. The well-fed parents are no doubt still in a cozy corner of the warehouse, starting another family by now.

MORE ON THE SUBJECT

For further reading the student is referred to the following sources described in the Bibliography, Appendix B.

Reference 7 *A Self-Inspection Program for Foodservice Operators* (NRA, 1973).
16 *Pest Prevention* (NRA).
25 *Protecting Our Food* (USDA, 1966).

PART FOUR
SANITATION
AND THE CUSTOMER

CHAPTER **11**

The 'Guestronomics'
Of Safe Food

In most industries you have to pay for surveys on what the customer thinks of your products and services. You the foodservice manager are lucky. Your surveys are free of charge, and if you are attentive you can get the results almost immediately.

Each time you welcome a customer into your establishment you are inviting him to survey the entire spectrum of your operation. You are asking him to tell you, directly or indirectly, what he thinks of the culinary quality of the food, the decor, the conduct of your personnel, and—most important for our purposes in this book—the sanitation standards achieved.

Unfortunately, many customers will not voice their complaints to you. Ask these timid souls how they liked the meal and, chances are, they will say they liked it fine, even if they despised every mouthful. If they are displeased, they will probably just leave and solemnly promise themselves never to return.

Even so, close observation of how customers behave can be highly suggestive for the foodservice manager. He can learn a lot by conscientiously trying to see his facility as the customer does. The manager's own eyes and ears—and even his nose—become the survey instruments.

It is of course important that the customer approve of the taste and appearance of food, and of the service he receives. But, you may ask, what does he know about sanitation? Since he sees only a part of the operation—the dining room and washrooms, at most—isn't his view a very narrow one? Assuming he did see the kitchen, what could he tell you about it that you don't already know? And if you ran an unsanitary kitchen which looked spanking clean, would he know the difference? Aside from keeping his patronage, why should you trust him as a guide to foodservice sanitation?

The customer may have an incomplete understanding of sanitation, but his behavior in reacting to your establishment can reveal a number of valuable things about its sanitary character.

That is the premise of this chapter, in which we urge you to examine the proposition that:

- **The customer usually has at least an intuitive sense of unsafe conditions in a foodservice, and that**

- **The manager can identify many shortcomings in sanitation by viewing his facility as a customer would.**

The Customer As Critic

However objective or subjective, the patron is the severest critic your foodservice can have. He brings a fresh view to your operation—if only because he spends less time there than you do—and he can quickly spot many of the defects you may have overlooked or become immune to by repeated exposure.

He is betting his health, pocketbook, and taste buds that you will give him a wholesome meal in clean, pleasant surroundings. He does not care about the time and effort that must be invested in foodservice sanitation, or about all the troubles which afflict a manager in achieving it. He very likely would care little that your work force had been decimated by the flu, that your dishroom supervisor had decided to pursue the arts, or that your meat freezer was on the fritz. He is not as ready as you are, in the thick of things, to make excuses. He is interested in results.

It is probably true that the customer has only a hazy understanding of the difference between "clean" and "sanitary", but his standing as a critic is not completely invalid. He does want things clean, and he prefers hot foods served hot and cold foods served cold. The technical differences between dirt-free and germ-free may not mean much to him, and the hazards of holding food in the temperature danger zone may hold no terrors for him. But hot soup is safer as well as more palatable. So we have to admit that the customer's judgment about unsafe food may be valid, even if he avoids it for the wrong reasons.

There need be no thought of deception in giving the customer what he wants. The manager, indeed, is foresworn to give him what he needs. We are concerned here with what is, in part, the public relations side of sanitation. Showing a clean dining room while giving short shrift to safe practices in the kitchen would of course be unthinkable, but the reverse would be sheer folly.

The Customer Acts

It is essential, then, that the customer have confidence in the foodservice, and that his confidence be confirmed on every visit. But how are you to tell that his confidence is waning or has vanished if he doesn't express his complaint? By closely—but discreetly!—watching him for the many unspoken signs of displeasure.

If you see a patron wiping tableware with a napkin, he may be one of a small group of people who do this as a nervous habit. Or it may be that the dishwasher is not doing its job, and the customer is trying to wipe away the visible evidence. Any time customers ask for tableware replacements—because a cup bears lipstick marks or bits of food are stuck to a fork—there is sufficient reason to check on dishwashing procedures. In this case, "clean" and "sanitary" are one and the same. If the souvenirs of a previous meal have not been removed from the serving ware, the same may be true of harmful bacteria which could easily have been deposited at the same time.

When a customer asks to be seated elsewhere because of a "draft," he may be overly sensitive to temperature changes, or the ventilation system may not be operating properly. Drafty conditions can raise dust and contaminate unprotected foods. One serious food poisoning incident was traced to bowls of pudding set out to cool in a windy, dusty corridor. Too *little* air movement can be just as bad since it may promote bacterial growth and insect infestation.

A plate of food left untouched can mean a number of things, and it may well be a sanitation danger signal. Possibly the customer's eyes were bigger than his appetite, or he wanted to experiment with a new dish and lost gustatory courage after it arrived. But it may also be that the food had an off-taste (indicating spoilage or worse) or that it included an unwelcome hair or insect fragment (indicating unsafe food handling). Similarly, the couple who ordered an expensive meal may rush through it because they are running late for the theater, or they may have found the surroundings so unappetizing they wanted to put the experience behind them as quickly as possible.

Dishes sent back to the kitchen are another sign that your sanitation guard may be down. When a customer asks his waiter to heat up the soup, that should tell you that safe, as well as savory, holding temperatures are not being maintained.

It is a curious trait of some patrons to reserve negative comments until they are at a safe distance. All through the meal they give you no reason to suspect that anything is amiss, perhaps because they find it too embarrassing to complain. They pay their bill and then, two paces from the door, in a voice meant to be overheard, tell a companion: "The food was okay, but their washroom is a sewerhole!" Or, "Our waiter looked like he had been swimming in the gravy." Or, "We ought to sue this place: I nearly tripped over the cockroaches!" The manager who usually takes station near the cash register might well move closer to the revolving doors for an uncensored critique of his operation.

Put Yourself In The Patron's Seat

If you have owned, managed, or or worked in a foodservice establishment for any length of time, you have probably become accustomed to its shortcomings. You no longer try to correct defects; you hardly notice them. But you will remain sensitive to its failings if you compel yourself to keep a customer's-eye-view. It is no easy psychological trick to put yourself in another man's shoes, but it will pay you to try. Play the game all the way. Tell yourself: "This is not my place. I have come here to get a meal, and they had better give me my money's worth." Forget about all the difficulties you are having with suppliers, the staff and the landlord. You are King Customer, and these are not your concerns.

Your impressions as a customer start even before you enter the establishment. If the parking lot, sidewalk, shrubbery, and entryway are free of litter, you will get a good feeling about the interior even before you see it. If the reverse is true, you may begin to doubt the cleanliness of the entire operation. As a manager-customer, you may also wonder whether the litter hides insect colonies and rodents waiting for an opportunity to enter.

Now sniff the air. Subtle cooking odors are acceptable, even appetizing. But an eye-tearing cloud of smoke and oily vapors from the exhaust vents are not. In addition to being hard on the neighbors and in violation of air pollution ordinances, accumulations of grease may provide breeding and feeding sites for insects.

Assume that all is well outdoors, so that when you step into the establishment your eyes, ears, and nostrils are fully sensitive to all the telltales, good and bad. Some of these indicators will relate more to esthetics than to sanitation, but you, the customer, will not make a conscious distinction. Grimy posters in the entryway do not bear directly on food safeness but they will have their effect on the customer. Dusty furnishings in the waiting room and cobweb veils over ceiling light fixtures, however, are patently unesthetic *and* unsanitary.

You enter the dining room. The first sound that assails your ears is the teeth-gritting noise of tableware being dumped into washbins. Such clatter hardly contributes to pleasant dining. It also does violence to the sanitary setting. Chipped, cracked dishes and glassware are difficult to clean and provide numerous tiny refuges for bacteria.

Still fresh from the outdoors, you detect an unpleasant odor only partly masked by food aromas. As a patron, you suspect decaying food particles and scraps which have collected in and around chairs, benches, booths, and tables—hardly the hallmark of a clean establishment. As a sanitarian, you also realize that bacteria may be the cause—and the effect. Many common microorganisms produce characteristic odors that are sharp, musty or pungent. In any case, conditions that produce unpleasant odors also provide breeding sites for insects and vermin, and they in turn can produce malodorous conditions.

Dark, damp, and uncleaned crevices in dining and service areas are highly vulnerable to insect invasion. The problem is especially acute around bus stations and service areas subject to frequent food spills. Bar and fountain areas are also susceptible because of high moisture conditions.

With the smell still in your nostrils, your first reaction may be to order someone to apply deodorant, disinfectant or pesticide in the suspect areas. Further reflection should tell you that covering up the odor or destroying bacteria and insects is only a temporary solution. Instead, get to the root of the problem. Why is food and excess moisture collecting? Perhaps because of poor work habits on the part of busy employees who splash and spill as they rush through their tasks. Or there may be a leak in a water or drain line. Possibly the facility itself is poorly designed, so that workers do not have enough traffic or storage space. An aisle may be too narrow to negotiate easily, forcing workers to squeeze past each other. The location of equipment and the design of furnishings may make them difficult to keep clean. A leg-mounted steam table may be so low to the floor that cleaning tools cannot easily be used beneath it. Heavily upholstered seats may defy the most vigorous brushing to remove bread crumb deposits.

All through the dining room there will be situations and events which cause you, the manager-turned-customer, to question the attention given by your staff to sanitation standards. Start with your own table: Is the linen freshly laundered and free of stains? (The customer may even be repelled by *clean* stains.) Is your menu card thumb-marked and greasy? Are the sugar, salt, and pepper containers sticky from too much handling? Are salad dressings, coffee cream and condiments left out on the table? (Many products may be safe without refrigeration until the container is opened and the contents exposed to the air.) Are the table and chairs and the floor clear of dust, litter and food scraps? Are water glasses and tableware streaked, spotted or caked with bits of food? (A fork that bears a speck of egg from the breakfast trade may be sanitary, but it isn't clean, and the customer is not enchanted by either.)

Next, survey the dining room. Dusty drapes, streaked or grimy windows, smudged or cracked walls, cobwebbed ceilings, loose floor tiles, or a heavily soiled carpet should make you feel a bit uneasy about the wholesomeness of the food. The customer might overlook those failings in his home, but he is much less tolerant when eating out.

Try to evaluate your personnel as the customer would. For him, the waiter and the busboy may be the only visible members of the foodservice staff. If the busboy has a clean uniform and is well groomed, you feel reassured. If he is slovenly and rude, you feel quite the opposite.

You spy a waiter taking a cigarette break in a corner of the dining room? As a patron you might worry about ashes in the food. As a sanitarian, you would also be concerned that he could transfer micro-organisms from mouth to cigarette, to hands, to food. If he coughs or sneezes over food, your reaction as a manager-sanitarian and as the customer pretty well coincide. Your confidence in the foodservice will similarly be shaken if the waiter touches food or food-contact surfaces.

Now it is time to examine one of the most customer-sensitive areas of a foodservice—the restroom. As the manager, you know that careless patrons often defeat the most sincere efforts to keep restrooms clean. As a customer, you may not be very indulgent about management's problems with other patrons. However endless and thankless the task may seem, restroom sanitation is an absolute must. Floors, walls, ceilings, fixtures and mirrors should be kept scrupulously clean. Handy and capacious receptacles will help keep the area clear of used paper towels and other litter. Good ventilation will combat the odor of disinfectant, and—better yet—the need for it.

The Moment of Truth

No matter how faithfully you adhere to the principles of sanitation, some sorry day the regrettable will probably happen: A customer will find a fly in his soup, a fragment of glass in his salad, or —heaven forbid—he will be the victim of foodborne illness. Such a moment of truth is fraught with manifold problems for the manager.

Any time a customer considers that you have imperiled his health, he can take his complaint to court. Your object is to avoid that problem, as well as to retain the customer's good will and patronage. And the first step in doing so is to realize that the customer may indeed have a legitimate grievance. Too many foodservice operators assume at the outset that the customer's complaint is unjustified.

If contaminants have found their way into a customer's food, or if the food has made him ill, you have an ethical and legal obligation to be concerned. Short of calling your lawyer, you must show that concern . . . to the customer. If he thinks he is being treated like a crank, he will be more inclined to press the issue.

Demonstrate a genuine interest in the customer's complaint and assure him that you and your staff make every effort to provide safe and wholesome food. Offer to obtain medical attention for the customer if the contaminant is one which could cause injury or illness. Your own sincerity and professional attitude will do much to relieve the customer's resentment.

Your next step should be to determine how the contamination occurred. An unfortunate incident can be a costly but valuable lesson. Don't ignore the lesson. How did the glass or fly get into the food? Perhaps an unprotected light fixture is located too close to food preparation areas, and a bulb was shattered in an accident. The back-of-the-house windows and doors may not be adequately screened to prevent entry of insects. Find the source of the mishap and you will be better able to prevent a repetition. Finally, don't be complacent because you get few complaints. Some customers are reluctant to tell you when something is wrong. They simply don't come back.

Even if a customer becomes ill after eating in your facility, he may be too far away when the symptoms strike to register a complaint. A recent case history is illustrative. A planeload of passengers were subjected to a long delay at a foreign airport. Dinnertime came and the airline took them to a restaurant at a railway terminal in a nearby city. Afterward the passengers re-embarked and several hours later, out over the Atlantic, a number of them became quite ill.

An airline representative investigating the unfortunate incident determined that the railway restaurant had been at fault. Off he went to the European city and confronted the restaurant operator, who was most indignant. "I've never had any complaints about the food", he insisted, "and no illnesses have ever been reported to me."

Technically speaking, he was right. Most of the restaurant's patrons were travelers, and any of them who became ill would be miles away by the time their illness became apparent.

So don't assume that all is well with your establishment just because no one has told you about his stomach upset after eating your food. Eternal vigilance is the price of high sanitary standards. It is also the price you must pay for peace of mind—the assurance that no one is going to become ill as a result of having been your guest.

IN SUMMARY

Each time you welcome a customer into your establishment, you are inviting the scrutiny of your severest critic—and, likely as not—a pretty competent judge.

Although many patrons may have an incomplete technical knowledge of sanitation, they may well have correct beliefs about it, and their reaction to standards of cleanliness and food-handling practices often point to very real problems.

Restaurant guests are often reluctant to voice a complaint but will reveal their displeasure in a number of unspoken ways. The observant manager can profit by learning to read the signs. He can also learn a great deal by putting himself in the customer's place, and judging his sanitation performance as the customer sees it.

The customer's viewpoint is valid because he brings a fresh and objective attitude and may observe defects you have overlooked because of over-familiarity with the operation.

Even the most carefully maintained foodservice facility may eventually experience a contamination incident. Faced with such an occurrence, assure the customer of your genuine concern and take advantage of the lessons learned to prevent a recurrence.

A CASE IN POINT

The customer was obviously a crank. That's what Tim said to himself, but he still tried hard to remain gracious as he responded to one demand after another. First, the customer complained about the tableware. The fork and knife were dirty, he said, even though they looked quite clean to Tim. And the dishes were greasy to the touch.

Then the customer expressed displeasure with the food. There were specks of dirt floating in the soup. And the meat tasted a bit sour. And the salad was gritty.

Tim quickly arranged for replacements and, out of sight of the customer, personally tested the rejected items. No matter how critical he tried to be, he found their flavor and cleanliness quite acceptable.

Obviously, Tim's guest was being picky. The complaints stopped as soon as Tim provided new tableware and food and showed concern. Maybe he just wanted to be pampered, Tim concluded.

The customer was a crank, or was he?

Quite possibly, but Tim shouldn't dismiss the matter quite so readily. Something very real may have caused the customer to question the sanitary standards of the entire foodservice. And it may not have had any direct connection with his food or table setting. Perhaps he saw something unappetizing — a roach, a slovenly foodhandler, food scraps, litter — which caused him to imagine that the meal served was also "dirty." It is good policy to give every customer complaint serious consideration — even when it hurts.

MORE ON THE SUBJECT

For further reading the student is referred to the following sources described in the Bibliography, Appendix B.

Reference 5 *Sanitary Techniques in Food Service* (Longrée/Blaker, 1971), Part III.
7 *A Self-Inspection Program for Foodservice Operators* (NRA, 1973).
9 *Profits and Your People* (NRA, 1972).
10 "The Sub-Standard Washroom" (NRA, 1966).
13 *Attitudes About Sanitation in Restaurants* (NRA, 1966).
23 *Food Service Sanitation Manual* (USPHS, 1962).

PART FIVE
SANITATION MANAGEMENT

CHAPTER 12

Employee Training

The foodservice manager must necessarily be interested in the "bottom line" of his profit and loss statement. Therein lies one key indicator of his success as a manager and his chances for the future. Further, he must know and understand all the factors which go into developing that bottom line — both income and expenditures.

Some of these factors are easy to measure and to record, like customer counts and check averages on one side, and like costs of products he uses (meat, poultry, fish, etc., — even including paper products and detergents) on the other. Other factors are not so easy to cost out. One of these is the cost of people, particularly non-management people. For knowing and understanding labor costs is much more than recording the payroll and adding the cost of fringe benefits and applicable taxes. The foodservice industry has some very special labor cost problems — sanitation training being one of them — which never appear as a line on the ledger sheet. Yet these problems are very important elements of the mana-ger's job, and the way he handles them may spell the difference between his success and failure.

This chapter will:

- **Discuss the role of training as a management function, including techniques and methodology,**

- **Indicate the benefits to be realized from a training program properly designed and implemented, and**

- **Consider training evaluation and the benefits of reinforcing employee performance which meets desired levels.**

The Training Gap

All foodservice managers and prospective managers (and unfortunately all too many customers) are aware that the industry has a higher rate of employee turnover than many other industries, and that it also has a history of severe difficulties in recruiting competent and motivated workers. The National Restaurant Association estimates that 225,000 new non-management employees must be hired each year of the next decade to keep pace with growth and replacement demands. Having employed approximately 3.7 million people in 1973, the industry is expected to pass the four million employee level before 1980, and to continue to grow! The expanding American eating-out pattern of the 1970's — influenced by the ever-increasing mobility of our society, a shorter work week, more women wage earners, and a steady rise in the general standard of living — is fully expected to continue. And this means that more people must be served good, wholesome food by personnel trained to the task.

At the individual restaurant or other foodservice operating level, where all non-school training ultimately must be done, the task of ensuring that all employees are fully conversant with the fundamentals of sanitation falls directly on the manager. Similarly, the task of ensuring that all employees consistently observe safe practices in performing their jobs falls directly on the manager. Needless to say, the manager's burden is not made lighter by high turnover rates, large numbers of new employees and the need to keep pace with ever-growing customer requirements. And even with experienced personnel who have not had formal instruction in sanitation methods, where does the manager start?

"The Magic Apron"

Sometimes he doesn't start at all. Or, more often, he assigns each new recruit, with or without orientation training, to the employee he is replacing or to some other person in the same work area. Thus, a new busboy would be assigned to "follow" another busboy or a waiter through his routines. A new cook's helper would be assigned to "follow" a cook. This "watch him," "one-on-one," or "magic apron" system so widely found in foodservice establishments seems almost a calculated effort to guarantee that errors of the past are continued into the future. The lowest common denominator takes over. Vital information, on subjects like sanitation, may unfortunately never be transmitted at all.

The "Crash" Method

On the other hand, the manager, possibly spurred on by a recent sanitation incident, an anticipated sanitary inspection, or personal concern, might undertake a "crash" training program. This kind of program could take any number of forms, some of them involving substantial expenditures of time and money for special training equipment and materials. Others may involve development of manuals for sanitation training, hiring of outside consultants, etc. But "crash" programs, while possibly useful for a time, are usually short-lived. The relatively continuous recruitment of new employees, perhaps one or two at a time; the facile but not necessarily accurate assumptions that one-time training is sufficient and that "experienced" workers require no further training — these are only some of the reasons. Indeed, it is vitally important in the first place for the developer of a "crash" program to recognize that it is just that, and to set about immediately organizing a long-range training program structured for all employee levels.

Year-Round Training

The answer, therefore, for today's competent foodservice manager, is the establishment of a training program as a company responsibility, keyed to the development of all his employees on a year-round basis. At the same time it should be designed and conducted so as to reassure his employees that management is interested in them, that their jobs are important, and that it is a personal (rather than a personnel) program. This fact should be emphasized by the designation of a training director or trainer who is involved in employee training *and* employee progress, and accountable for both. And there must be a budgeted commitment of company resources (time, money, men and materials) for this training.

"Is this a trainer's theory or an author's pipedream?" you might well ask at this point. The response is emphatically "No!" This very thing is happening today even in progressive companies in the foodservice industry. Recent surveys reveal annually increased training budgets, some now ranging to several million dollars in larger companies, and in excess of one per cent of gross sales in medium-size companies. Further, according to the magazine *Institutions / Volume Feeding Management,* the most frequently established new executive position in foodservice companies in 1973 was that of Training Director. In fact, industry training directors have even established their own professional association, The Council of Hotel and Restaurant Trainers (CHART) — and sanitation training has been high on their conference agenda.

Yet, in restaurants and other foodservice establishments professional restaurateurs rather than professional trainers must normally "do" the sanitation training. An interested member or representative of management should personally conduct the course, thus providing a real and visible commitment to the program. Employee committees can be used to enhance personal involvement on the part of workers — meeting, where effectively utilized, the organizational and training needs of both the employees and the company. In addition, the continuous nature of the program can be emphasized.

ESTABLISHING OBJECTIVES

Establishment of instructional objectives, preferably in a behavioral format which clearly states *what the trainee will be able to do* on completion of training, is a first step in any training process. It is critical to successful sanitation training, since employee performance is the purpose and the measure of the training. Further, by tying training to performance, it is possible to determine why and what training is needed, and for whom. Each employee in a foodservice operation should receive sanitation training to protect the public as efficiently and effectively as possible. But it would be both wasteful and foolish to train at too technical a level or to provide specific training for those who will never use it.

Objectives should be set, therefore, for each job category in the establishment, while recognizing the common content required for performance of all jobs. This content should cover, at a minimum:

1. Personal sanitary practices which every employee should follow during work and before coming to work.

2. Food protection, including information on proper foodhandling procedures and the dire consequences of not following these procedures.

3. Sanitation of facilities and equipment, including special cleaning and sanitizing required in particular areas.

4. Pest control, particularly the practices necessary to eliminate cockroaches, flies and other insects, as well as rats, mice and other vermin.

New employees and trainees, whatever their intended job, can well be considered as a separate job category in setting objectives for sanitation training. The fundamentals included in the above common content should be stressed for all new employees, preferably before they undertake job assignments.

TRAINING MATERIALS

It may be possible for an especially gifted instructor to use words alone to describe the operation of a dishwasher, the setting of a table or the fundamentals of foodservice sanitation. However, the mere fact that words have been spoken certainly does not guarantee that anything has been learned by his student or employees. Even words that land their message may not be enough. Instructional aids of all varieties, ranging from drawings, charts and graphs to mock-ups and films, raise the odds considerably in

favor of training effectiveness. If they are supported by realistic trainee participation and actual exercises and on-the-job experience, the odds go up even higher.

Proper use of training materials saves time, adds interest, helps trainees to learn, and, of course, makes the trainer's job easier.

In the choice of materials for sanitation training, the manager or trainer should be guided by four "A's." To be most useful, training materials must be:

1. Accurate — that is, factual, up-to-date, complete, and in an understandable form.

2. Appropriate for the purpose they are to serve, and, most importantly, for the trainees who are to use them. Language levels, reading levels and difficulty levels must be considered and matched.

3. Attractive. All teachers, and probably all parents, know that one cannot teach anything to a person without first gaining his attention. In subjects which do not generally have a wide appeal (and sanitation is one of these), it is often necessary to go to great lengths to make information exciting and interesting and — what is often most difficult — to make it memorable.

4. Authenticated — that is, they must bear the seal of an identifiable authority in the field.

No one set of training media can do the job effectively for all trainees at all levels. Many companies have developed employee handbooks which are distributed to new workers and from which workers are expected to glean necessary information. They are far from adequate for sanitation training but can be useful in other aspects of orientation and as a guide to company training.

Audio-visual programs and aids, ranging from motion pictures to sound-film-strips and slides, are increasingly used in the industry. However, the trainer cannot expect an audio-visual presentation to be a total training program or the learner to be motivated solely by a film. Rather, the trainer should introduce each showing with a statement of objectives and expectations, and discuss the presentation's content as it applies to his restaurant or company, as well as to the employees involved in the training. Practical and realistic demonstrations should be stressed. Finally, a short examination such as those in the National Restaurant Association publication *Profits and Your People* should be given, scored by the trainee or trainer, and indicated problem areas clarified through discussion. Signs, posters, bulletin boards, and pay envelope stuffers can serve effectively as reminders or refreshers associated with sanitation training and proper sanitation practices. However, they are not in themselves a training program or a substitute for individual and personal training.

SCHEDULING TRAINING

The difficulty of finding a time and place for training in a busy foodservice establishment is often given as an excuse for the lack of a training program. Obviously special steps must be taken (and can be taken) if management is truly committed to training, and cognizant of the benefits which can be derived from an effective and scheduled program. We all know that wasteful and inefficient employees cost money. Consider,

then, the irredeemable damage that can be done to a foodservice operation by an employee who is untrained in the fundamentals of sanitation. Consider the impact on a manager's career of a salmonella outbreak traced to his restaurant or hospital foodservice — and announced on radio and television. Or consider the chances of repeat business from customers who have seen cockroaches in the soup or rats near the back door.

Scheduling sanitation training for foodservice employees is essentially a matter of deciding priorities. Who should be trained first? Should it be a new employee, or an employee being given new responsibilities? Should it be a group of employees, such as waitresses who have never received formal training in sanitation? Or should it be all members of the staff, to call renewed attention to general health and safety rules?

Probably it should be all of these, and a master training program with blocked-out time scheduling can be useful in showing both the company's and manager's commitment and the application of the program to all employees.

Special attention can be given to priority situations — a large group of new employees, an opening of a new restaurant, a re-opening after remodeling, or special preparations for banquet business or for a convention group. But adaptations such as these should only be made when a master plan has been developed.

Training sessions should not be too long, probably never longer than 30 minutes, and should be announced well in advance. Notices should indicate the date, time and place of each session, the subject, who is to receive the training and who will conduct the session. The manager should make sure that announcements are posted where they will be readily seen, and that everyone concerned is aware of management's commitment and that company resources are involved. The trainees should also know that each session will begin on time and that unexcused absences will not be tolerated.

CONDUCTING TRAINING

Training should be conducted in an atmosphere in which employees feel at ease. There is normally no need for special training rooms, although they should be used when available. Rather, in most foodservice operations, a place should be selected which is casual, friendly and conducive to open discussion and the exchange of information. The area need not be large unless the group is large. An employee lounge, executive office, part of a storage room or a separate section of the dining room not in use can serve adequately.

The training area should, of course, have adequate and comfortable seating, clear writing surfaces, and electrical outlets for any training equipment to be used.

There are many learned texts on how to instruct, and sanitation training is not unlike many other areas of teaching in this regard. The foodservice sanitation instructor should be thoroughly familiar with the content of his training program and the subject of each individual session — and he must prepare in advance. The foodservice employee can spot an unprepared instructor as easily

as any other student can.

But there are differences from standard classroom teaching in any program where ability to perform, rather than recitation of knowledge, is the measure of success. The training program is only as good as the trainer, and the trainer only as good as the preparation he has made for conducting training sessions. The successful trainer familiarizes himself with the problems, equipment, methods and procedures which confront the trainees in their daily work. He studies the trainees and, where possible, predetermines ways of arousing and maintaining maximum interest and enthusiasm. He also anticipates situations and problems which may arise. He understands that the keys to his success are his relationship to the trainees and the relationship of the training to actual jobs. To generalize, it is probably accurate to say that there is no such thing for a foodservice worker as "sanitation" which is not directly related to his work.

Training in sanitation must therefore take the form of explaining how specific tasks are carried out to serve sanitation goals. The actual method of training can vary. However, if there is one element common to all effective training programs, it is participation by the trainee. Whether it is some form of "learning by doing" (explain, demonstrate, perform, correct, re-perform, etc.), conferences, role playing, or dialog role playing, getting trainees active in self-training is essential to success. It will be the trainer's responsibility to help individuals develop their own awareness, sensitivities and habits of approach to their jobs and the tasks which make up these jobs.

If possible, a training team arrangement should be utilized, soliciting key employees to assist in training of new personnel, personnel moving to new assignments, and those experienced personnel requiring refresher training. Key employee participation should be recognized through appropriate title designations, and reinforced by that most tangible inducement — financial reward. All evidence indicates there is no more real or visible manifestation of a company's commitment to training than specific remuneration for line employees serving additionally as trainers.

Each trainee should also receive some recognition when he has completed a training program, preferably as soon as possible after the training. Awards and certificates are commonly used, but the form of recognition should be matched to the situation. In any case it should come from management, and its presentation should clearly indicate that the manager is genuinely interested in the trainee's progress.

MEASURING PROGRESS

In any kind of training, objective measurement of individual progress is a valuable tool. Specific written tests or quizzes, such as those in NRA's *Profits and Your People,* can aid in determining employee knowledge of the content of training sessions. Oral questioning and discussion can also be useful both in the training sessions and afterward.

But since it is performance rather than ability to recite technical knowledge about sanitation which is our primary goal, the ability of foodservce employees to do their assigned tasks in the work situation is the true measure of progress. Where standards of accomplishment have been set before training, progress can be seen in terms of achievement with respect to applicable standards and measured against quantifiable objectives. The foodservice manager must use great ingenuity in customizing his employee evaluation system, using objective measures available to him, personal observation, and self-inspection both by key employees and himself. Employee turnover data, absenteeism and tardiness reports as well as productivity information have been found valuable in measuring progress of overall training programs. Customer surveys, complaint reports and customer return rate data can often be directly tied to the quality (or lack of quality) of sanitation training.

IN SUMMARY

One of the foodservice manager's most important responsibilities to his operation, his employees and his guests is training employees in the principles and practices of foodservice sanitation. Since foodservice in the broadest sense is a public industry, only through effective training can the manager protect the public from foodborne illness.

The special "people problems" of the industry — arising from growth demands and patterns, high employee turnover rates and the fact that most training must be done at the level of the individual restaurant or operation — make sanitation training a difficult and yet a continuous need. Year-round training programs on an individual basis for both new and experienced employees are essentially a public service requirement placed on the foodservice industry.

The benefits accruing from a properly designed and conducted training program are demonstrable in very practical and measurable terms — dollars. Progressive restaurants and foodservice companies are recognizing these benefits on an increasing scale. Full-time or part-time training director positions are being established in practically all major multi-unit companies, in many medium-size companies and in more and more smaller companies interested in future growth. Food service sanitation training must be a common denominator in the job descriptions of all these positions.

A CASE IN POINT

One otherwise ordinary Thursday afternoon, Tim received an unexpected visit from the Health Department. The visit was a surprise because the sanitarian had conducted his regular inspection only a week before and, except for minor discrepancies, gave Tim's restaurant a clean bill of health.

The news today was a different story. Several residents of the city had become ill and reported eating in Tim's restaurant. It looked like trouble. Whatever the outcome, they were Tim's customers and the restaurant's reputation was in jeopardy. It was agreed that no case of food poisoning had ever been traced to his place before. But *action* to determine the cause of these cases, and head off any further outbreak, must be initiated immediately.

Tim and the sanitarian got down to business. Test samples from possible contamination sites were taken to the laboratory, with the following results: A roast beef and gravy entree served earlier in the week was identified as the source of the disease germs implicated.

It *had* happened here. Personnel error was involved in what was clearly a serious breakdown in the restaurant's food protection program.

After taking steps to rectify the immediate situation, what should the manager do to prevent a recurrence?

There are a lot of answers to this question: proper supervision, careful monitoring of time and temperature as applied to food on the serving line, and personal hygiene, to mention a few. But the long-term answer is training. A year-round training program in foodservice sanitation for all employees must be considered an indispensable follow-up to Tim's immediate action plan. Nothing less than qualified personnel can ensure that only safe food will be served.

MORE ON THE SUBJECT

For further reading the student is referred to the following sources described in the Bibliography, Appendix B.

Reference 7 *A Self-Inspection Program for Foodservice Operators* (NRA, 1973). See digest in Appendix A.
8 *Career Ladders in the Foodservice Industry* (NRA, 1971).
9 *Profits and Your People* (NRA, 1972).
13 "Attitudes About Sanitation in Restaurants" (NRA, 1966).
14 "Sanitation Checklists for Management" (*Cooking for Profit,* March 1972).
35 *The Preparation of Occupational Instructors* (HEW, 1966).
37 *Preparing Instructional Objectives* (Mager, 1964).
38 *Sanitary Food Service, Instructor's Guide* (HEW, 1969).

CHAPTER 13

Regulations and Standards

If at this point you are a bit overawed by the demands of foodservice sanitation, be reassured. You are most certainly not alone in the battle against food contaminants. There are a great many people and organizations who are trying to help and guide you. In fact, some of them will insist that you take their advice!

First there is government, at various levels, which suggests or prescribes rules for you to follow, and inspects your operation to make sure you are complying with the compulsory provisions of law. Next are the numerous professional and industry groups which set standards and offer guidelines for safe foodhandling, proper equipment design and other measures calculated to prevent foodborne illness.

Every facet of a foodservice is inspected, evaluated and regulated in one way or other by government and industry organizations. Some foodservice operators, indeed, view much of this interest in their affairs as unwarranted intrusion, but that attitude may be excessively narrow. No matter how capable the operator, the task of providing safe and wholesome food, at competitive prices, is a large order — and too much depends on it.

A more viable attitude is to recognize that constructive guidance is available, and to consider how to realize the greatest benefit, public and private, from it. It may seem burdensome to have inspectors scrutinizing every area of your operation, but if you are receptive and cooperative you very likely will find them offering practical solutions to some of your most vexing problems. With this approach the operator will appreciate the value of regulatory and standards-setting agencies and resolutely seek an understanding of how they can help him. We will briefly examine the two fundamental guidance and control systems at work in our society to protect the sanitary quality of food:

● The official system of regulatory and advisory controls administered by agencies of government.

● The unofficial system of voluntary controls conducted by private industry, trade associations and professional groups.

Marry these two and a highly effective complex of statutory and self-imposed standards and regulations can be seen as a very real possibility.

GOVERNMENT REGULATION OF THE FOOD INDUSTRY

So important is foodservice sanitation to public health that government has long since established regulations and standards and provided for their enforcement. The first health ordinances in the Western Hemisphere were adopted more than 400 years ago, and the nation's first "state health officer" was appointed when the United States was still a group of English holdings.

Government control is exercised at three jurisdictional levels: federal, state and county/municipal. Typically, the foodservice operator is not directly affected by all federal laws regulating food enterprises, but federal standards for foodservice sanitation are generally reflected in state law and local regulations. U. S. regulations do, however, safeguard the sanitary quality of many food products before they reach the operator.

Federal controls of greatest significance for the foodservice manager are those administered by the Food and Drug Administration of the Public Health Service, a major entity of the U. S. Department of Health, Education and Welfare. The Food Service Division of the FDA Office of Nutrition and Consumer Sciences formulates and promotes the acceptance of a recommended ordinance and code for foodservice sanitation — the model followed in whole or in part by most regulatory bodies in the United States. This document is periodically revised to keep it abreast of developments affecting the industry.

The FDA Office of Compliance enforces mandatory provisions of the U. S. code regulating foodservice in interstate carriers — air, rail and ship. This office also performs an over-all regulatory function with respect to food products in interstate commerce, including inspection of food processing plants, to ensure adherence to standards of purity and wholesomeness and compliance with labeling requirements. The FDA Office of Technology has cognizance over milk sanitation standards and standards for the shellfish industry, including the water quality of fisheries and safe handling methods in processing and distribution.

U. S. Department of Agriculture activities of primary interest to a foodservice are those associated with inspection and grading of meat, meat products, poultry, dairy products, eggs and egg products, and fruits and vegetables shipped interstate, as described in Chapter 5.

The U. S. Center for Disease Control, a field agency of the Public Health Service in Atlanta, Georgia, investigates foodborne illness outbreaks, studies the causes and control of disease, publishes statistical data and provides extensive educational services in the field of sanitation.

Other federal programs are significant but have a less direct bearing on foodservice operations. The Environmental Protection Agency sets standards for air and water quality, the use of pesticides and the control of wastes. The Bureau of Commercial Fisheries of the U. S. Department of the Interior develops noncompulsory standards for the sanitary quality of fishing waters and for safe processing methods. The Occupational Safety and Health Administration of the U. S. Department of Labor protects employees of foodservice establishments.

In the everyday operation of a foodservice, however, the laws which affect the operator most meaningfully are those enforced by state and local health authorities. These are the agencies which spell out the regulations and conduct inspections to ensure compliance.

From state to state and from city to city the regulations vary in language, coverage and manner of enforcement. The organization and scope of the monitoring agency will vary also, depending on the locality. In a large city, the enforcement agency will probably be a local one. In a small municipality and in rural areas, a county or state health department will have jurisdiction.

In any case, the manager has a duty to familiarize himself with the applicable laws and with the enforcement system in effect. Some health departments issue guidelines on foodservice sanitation that explain the law in simple layman's terms.

Regardless of differences in wording, the intent of regulations in any locality is essentially the same: To protect the dining public from foodborne illness. By and large, the mission of the public health officer and of the sanitation-minded foodservice manager are in harmony. They both want to protect people by protecting food from contamination, and by preventing the growth of micro-organisms that may succeed in getting into food.

As noted, most state and local ordinances are patterned after the model ordinance and code recommended by the federal government. An outline of the main provisions of the code will illustrate the range of government cognizance:

- **Food: Supplies and protection.**

- **Personnel: Health, disease control, cleanliness.**

- **Food Equipment and Utensils: Sanitary design, construction and installation cleanliness.**

- **Sanitary Facilities and Controls: Water supply, sewage disposal, plumbing, toilet facilities, handwashing facilities, garbage and rubbish disposal, vermin control.**

- **Other Facilities and Operations: Floors, walls and ceilings; lighting; ventilation; dressing rooms and lockers; housekeeping.**

- **Enforcement Provisions: Permits, inspection of foodservice establishments; examination and condemnation of food; foodservice establishments outside the jurisdiction of health authorities; review of future construction plans; procedure when infection is suspected; penalties.**

FOODSERVICE INSPECTIONS

To you, the foodservice manager, the law is most directly represented by a man with a clipboard — the health officer or inspector. In most communities, he must qualify for his job by education and experience. While he may have a somewhat different point of view, and talk a slightly different language, he is a professional in his field and carries the badge of authority.

The inspection process may begin even before a foodservice is established. Many jurisdictions provide for advance review of plans and specifications for new construction or extensive remodeling. This procedure makes sure that applicable codes will be met, benefitting the manager as well as the community. Obviously, it will delay the opening and prove expensive if an error has to be corrected after construction is finished.

Once the foodservice building is completed, an "opening" inspection visit can be expected prior to issuance of a permit, and several agencies may be involved. For example, if liquor is to be served, some states require inspections by both the liquor control and health authorities. If the establishment is to process foods for interstate shipment, the U.S. Food and Drug Administration and the U.S. Department of Agriculture become involved in the authorization procedure.

After a foodservice opens, inspections can be anticipated anywhere from once a month to once a year. The exact frequency will be influenced by such factors as the work load of the local agency and the severity of the violations found on previous inspections. Usually, the health officer will make an inspection report which must be signed by the owner, his manager or a designated representative, and the foodservice will be given a copy. The report is not just another piece of bureaucratic paperwork to be forgotten and discarded. It should be closely studied by the manager, and any violations noted should be discussed in detail with the inspector. The intelligent manager will want to know the exact nature of the violation; he will also realize that the inspector sees many foodservices and is in a position to recommend corrective procedures based on experience. Indeed, we would submit that the inspector who merely "gigs" but does not suggest solutions is hardly achieving effective enforcement of the law and protection of the consumer.

In any case, it is essential that violations be corrected promptly. Failure to do so may make the owner subject to a fine, unfavorable publicity, and, in extreme instances, closure of the establishment.

What is it the health inspector will be looking for? Which defects will he tend to consider major, and which ones will he think minor? The extent of inspection coverage, and the areas of greatest and least emphasis will vary between localities and even between individual inspectors. Some officers are most concerned with refrigeration temperatures; others may concentrate on the general physical appearance of the facility. One inspector may examine dishwashing procedures in detail, and another may give primary attention to the personal hygiene of workers.

The National Restaurant Association has found that health departments generally place major emphasis on the following areas:

— Food Protection: Food from approved sources, safe water supply. Safe storage, preparation, display and serving; safe temperatures applied. Foods protected from contamination.

— Personnel: Employees free of disease and infection. Personal cleanliness.

— Sanitation: Properly designed, constructed and installed facilities and equipment. Cleanliness of premises. Cleanliness of utensils and equipment.

— Chemical Hazards: Safe storage and use of toxic and other hazardous materials. Knowledge of dangerous chemicals to be avoided.

— Disposal of wastes; vermin control.

Rating the Inspector

But how do we rate the raters? That seems to be a universal question in any proficiency rating system. The U. S. Public Health Service provides an answer in its survey program for certifying state sanitarians. And many local health departments base their inspection coverage on the evaluation checklist used in the federal survey, which assigns demerits for failure in 118 specified inspection areas. These penalties range from 1 for the least discrepancy to 6 for the greatest discrepancy in an inspectors's proficiency. The numbers indicate, in a negative but interesting way, the value placed on various aspects of foodservice sanitation. The checklist is too extensive to reproduce in full, but it will be instructive to see how inspection proficiency in the "Food Protection and Supplies" section is rated. In the following list, the number alongside each item represents the demerits assigned for failure in that inspection routine:

Approved source 6
Wholesome, not adulterated 6
Not misbranded 2
Original container, properly
identified, or in approved dispenser 2
Fluid milk and fluid milk
products pasteurized 6
Protected from contamination 4
Adequate facilities for maintaining food at hot or cold
temperatures 2

Suitable thermometers properly
located 2
Perishable food at proper
temperatures 2
Potentially hazardous food 6
Frozen food kept frozen,
properly thawed 2
Handling of food minimized by
use of suitable utensils 4
Foods cooked to proper
temperatures 6
Fruits and vegetables washed
thoroughly 2
Containers of food stored off floor .. 2
No wet storage of packaged food 2

Inspect Yourself

It is far better, of course, for a manager to know that his place is sanitary than for him to worry about what the health officer will look for. The choice here is between positive *action* and negative *apprehension,* and the self-reliant manager will surely choose the former. Seen in this light, the best preparation for an inspection is self-inspection — by the foodservice operator himself — followed by corrective action as required.

Detailed self-inspection procedures have been developed by the Public Health and Safety Committee of the National Restaurant Association. These procedures are presented in a recent NRA publication, *A Self-Inspection Program for Foodservice Operators,* which contains a complete series of inspection routines in convenient checklist form. (Appendix A of this book reproduces these self-inspection procedures in abbreviated format.) By conscientious use of these checklists the manager will be able to anticipate the most comprehensive health inspection.

In the final analysis the success of standards, laws and enforcement programs will depend on how well they recognize the practicalities of foodservice operations. The foodservice industry can most effectively protect the consumer, with least disruption and cost, by working in close cooperation with government agencies. It is for this reason that the NRA Public Health and Safety Committee has urged establishment of joint food protection councils in state and local health jurisdictions. A number of these councils are already in being, actively cooperating with health departments in formulating regulations, operating training programs, and solving enforcement problems.

VOLUNTARY CONTROLS WITHIN THE INDUSTRY

Probably few segments of American industry have devoted as much effort to self-policing as have food, foodservice and associated enterprises. Scientific and professional societies, manufacturing and marketing firms and trade associations have energetically pursued programs designed to raise industry standards through research, education and cooperation with government. The overall result, as it pertains to sanitation, has been to:

— Increase understanding of foodborne illness and its prevention.
— Improve the design of equipment and facilities for greater cleanability, effectiveness and reliability.
— Help maintain the safeness of foods during processing, shipment, storage and service.

— Help make foodservice laws uniform throughout the nation, and give them practical applicability.

While an almost endless number of organizations have contributed to these improvements, those having the greatest relevance for the foodservice operator are:

The *American Public Health Association,* an umbrella organization whose membership represents virtually all individuals involved in any aspect of public health. Its members are doctors, nurses and sanitarians. The Food Protection Committee of the Environmental Section of the APHA serves as a forum for professionals in this field, establishing policies and standards for food sanitation on a broad base.

The *Association of State and Territorial Health Officers,* and the *Association of State and Territorial Directors of Local Health Services.* Members of both of these professional organizations are medical doctors who are public health officials. They exchange information on public health problems and preventive measures.

The *Conference of State Sanitary Engineers* and the *Conference of Local Environmental Health Administrators.* The membership of these bodies are sanitary engineers primarily concerned with the engineering aspects of sanitation, as in the design of water supply and waste disposal systems.

The *National Environmental Health Association,* an organization of environmentalists and sanitarians, including those responsible for food inspection services and for establishing sanitation programs; the *International Association of*

Milk, Food, and Environmental Sanitarians, pioneer in the highly successful U. S. milk sanitation program, and the *National Society of Professional Sanitarians.* These organizations are composed of sanitarians concerned with sanitation and food protection in national, state and local jurisdictions. They promote professional standards, recommend legislative policy and sponsor uniform enforcement procedures.

The *National Sanitation Foundation,* a non-profit research and testing organization which evaluates foodservice equipment and materials. The foundation has established widely accepted standards for equipment design, construction and installation. The NSF seal appears on equipment meeting those standards.

The *National Pest Control Association,* an organization of pest control operators which provides guidelines for pest prevention programs.

The *Association of Food and Drug Officials of the United States.* This body develops food sanitation codes and fosters consumer food protection through uniform legislation and enforcement. One of its major accomplishments has been a model code for handling frozen foods.

The *Frozen Food Industry Coordinating Committee.* This committee has developed a code covering frozen food handling from processing to retail foodservice.

The *Food Research Institute,* a non-profit organization which conducts research on the causes and prevention of food contamination and foodborne illness. The FRI works to benefit both food processors and foodservice operators by setting maximum permissible levels of micro-organisms, chemicals and chemical additives in food.

The *National Restaurant Association,* the national trade association for retail and institutional foodservice establishments. Through its Public Health and Safety Committee, the NRA promotes food sanitation in cooperation with governmental, scientific, commercial and educational institutions.

The kind of progress which can be achieved through these voluntary programs of the industry is exemplified by the model procedures for handling frozen foods as described in Chapter 5 (Food Procurement and Storage). Another example is the National Sanitation Foundation's approach to determining the suitability of specific units of equipment for use in a foodservice. Included in NSF evaluations are: accessibility of equipment for inspection and cleaning; corrosion resistance of materials, including non-transference of color, odor and taste to food; safeness of materials used for food contact, splash contact and non-food contact surfaces; integrity of welded seams and soldered joints; effectiveness of gasketing and soundproofing materials; safe properties of plastic resin compounds; safe radii for internal and external corners and edges; safety of fastening methods, exposed edging and nosing, reinforcing and framing; depth of retaining grooves (for cleaning ease) ; design of shelving; clearance to floor and design of vertical support members; weight factors for units intended to be portable; and function of controls.

IN SUMMARY

The foodservice manager is advised, guided and regulated in every phase of his operation.

Because of the importance of safe food to the community, government has long regulated the operation of foodservice establishments.

Industry organizations study his business activities, research his problems and recommend performance criteria, including sanitation standards.

A number of professional and trade associations seek to improve foodservice operating practices through education in sanitation, development of standards for foodservice equipment and facilities and for protection of food in processing and storage, and through research into the causes and prevention of foodborne illness.

Government at all levels recommends standards, and administers controls that directly or indirectly affect the foodservice operator. While federal regulations do not all apply directly to retail foodservices, state and local regulations which do apply are often patterned after the model ordinance and code recommend-ed by the Food and Drug Administration. Federal authorities also regulate the purity and safeness of foods in interstate commerce.

The language of foodservice laws may vary from state to state, but applicable regulations, whatever the locality, all contain provisions governing food safeness, personal hygiene, sanitary facilities, equipment and utensils, safe operating practices, and enforcement procedures.

Insofar as the individual manager is concerned, the most visible representative of the law is the local health inspector. A highly effective way for the foodservice manager to be prepared for visits from the health department is to institute a rigorous self-inspection program, for which competent guidance is readily available (see Appendix A).

Local laws and enforcement programs are most effective when there is close cooperation between foodservice operators and regulatory agencies. That is more than a better-government, better-business, better-community exhortation: it is a demonstrated, statistical fact.

A CASE IN POINT

At last report we left Tim in the most serious predicament of his career as a foodservice manager — an outbreak of food poisoning traced to his restaurant.

As we might expect, he has recovered from the shock and is busy organizing a training program and doing a number of other things aimed at warding off future troubles. Like self-inspection.

At an earlier time Tim might have treated a sanitation checkoff list as a means of getting ready for the public health officer. But now inspection items have become more than words on a checksheet. They are real things you want to do —to protect food and the people who eat it. Far from trying to get the jump on an inspector, he would welcome the most searching inspection that turned up all manner of embarrassing discrepancies. Anything to avoid another illness incident.

With characteristic enthusiasm, Tim went all out for self-inspection, making the rounds himself every day, checking every operation personally. By week's end he was nearly exhausted. A routine event shifted his attention to training and he realized he had almost forgotten about it. Then suddenly he saw the light. Inspection and training. He put the two together, and the answer was clear.

Would you care to anticipate the moral to this story?

Inspection is only as good as its follow-up, and Tim can't do it all. That means continuous training and supervision. And trained supervisors will give us what we are ultimately striking for — *sanitation at the source.*

MORE ON THE SUBJECT

For further reading, the student is referred to the following sources described in the Bibliography, Appendix B.

Reference 7 A *Self-Inspection Program for Foodservice Operators* (NRA, 1973).
11 "Know Your Health Officer" (NRA).
13 "Attitudes About Sanitation in Restaurants" (NRA, 1966).
14 "Sanitation Checklists for Management" (*Cooking for Profit*, 1972).
23 *Food Service Sanitation Manual* (USPHS, 1962).
24 *Current Concepts in Food Protection* (USPHS/FDA), "Legal aspects".
25 *Protecting Our Food* (USDA, 1966), "Government and Industry Roles".

CHAPTER **14**

Managing a Safe Foodservice

When it comes to food sanitation, the manager works in a hostile world indeed. Everywhere he looks, opportunities abound for contamination. The food he buys, the people he hires to prepare and serve it, the people who consume it, the very four walls (and floor!) of his establishment — all are sources of trouble.

Yet the foodservice manager must overcome these threats because he has the moral, professional and legal responsibility to manage a safe operation. Regardless of the type or size of his facility, he knows that the most damning indictment his operation could receive would be the report: "I got sick there once." The highest quality food, the most elaborate menu, the most gifted chefs, and the most gracious service would count for nothing in the face of that complaint. One stomachache could do much to destroy a reputation built up over the years.

The difficulties may seem overwhelming, but they will not daunt the manager who understands the problem and is prepared to deal with it. He has reduced all the complex factors to a simple concept: Food sanitation is a matter of *clean people* serving *clean food* in *clean places.*

Viewed in this manner, it can be seen that sanitation is best practiced as a natural, automatic part of every phase of the foodservice operation.

In effect, this is the approach to safe and wholesome foodservice we have encouraged throughout this book. Each word, each paragraph, each chapter has in one way or other been related to one or more of the three major components of sanitation: People, food, places.

Clean People

Safe food-handling practices should be an important consideration in the selection, placement and training of every worker. (And no person in the establishment should be better informed about sanitation — and more observant of the requirements — than the boss.) Starting with the first interview, prospective employees should be screened for their health and personal hygiene habits, as well as for the knowledge and abilities needed for their jobs.

House rules and policies on safe food-handling should be clearly stated on an employee's first workday, and then reinforced with regularly scheduled training sessions. Food sanitation should be an integral part of any program to maintain and improve the performance of a foodservice operation.

Nor can the patron be overlooked as a potential hazard. The establishment must protect itself and the public against possible infection *by* the public. Managers should be particularly alert to the danger of public contamination in self-service areas such as salad and appetizer bars.

Clean Food

All food, no matter how reliable the source, is naturally contaminated to a certain extent even before it arrives in the foodservice establishment. It is management's job to make certain that the facility presents a controlled environment in which contaminating micro-organisms can't pursue their activities unchecked.

For example, even freshly slaughtered meat is seldom free of bacteria. By the time it is ready for cooking, it may contain thousands if not millions of organisms per gram. Fortunately, the kinds of bacteria which cause most foodborne illness can be controlled through proper storage, planning and preparation.

Although the sources of contamination are various, the varieties of pathogenic organisms are relatively few. The most common bacterial foodborne illnesses transmitted to man through meat (including poultry and fish) are Staphylococcus poisoning, Clostridium perfringens poisoning, Salmonellosis, Shigella dysentery, and Streptococcus infection. In 1971, 89.8% of the individual cases of foodborne illness were caused by bacteria. Of this total, 38% were due to Staphylococci, 28.7% to C. perfringens, 6.7% to Shigella, and 5.6% to Salmonella. If only these micro-organisms were successfully combatted, the problem of foodborne illness would obviously be kept to a tolerable level.

It is not realistic to hope that all foodborne illness will be conquered. The products, practices and processes encountered in the foodservice operation leave too many opportunities for contamination. And it's not practicable to maintain absolutely sterile conditions — not so long as humans or their nutrients are around. But it *is* possible to keep things "sanitized clean" at all times, and that's what we need to operate a safe foodservice. The knowledge and technology exist to control foodborne illness, but it is up to management to apply them.

How we do that during the preparation and service of food — the times of peak danger — is the challenge. A few simple rules apply:

— Use bare hands only when absolutely necessary. Food should be handled with the proper utensils, properly cleaned. To do otherwise is to increase the chances for contamination and cross-contamination. Workers should be trained to recognize the fact that their bare hands are dangerous, and they should have the cleaning facilities and tools to keep that danger to a minimum.

— Keep hot foods hot, cold foods cold — never anything in between. Most foodborne illnesses occur because food has been held at temperatures which encourage the growth of micro-organisms. Sauces, dressings and gravies are left out at room temperature or over a low flame at considerable peril. Unrefrigerated sandwich spreads and cold meats can develop the properties of a booby trap.

— Bring every part of a food mass to a safe temperature as rapidly as possible. If food is heated or cooled in too large a mass, it is possible that the interior of the food won't reach the proper temperature to control microbial action. The result is that food may become hazardous even though it has been placed on the range or in the refrigerator.

— Don't prepare foods in advance of serving time without strict regard for time and temperature factors. Frozen foods should be thawed as directed. Fresh foods should be cooked or processed as required. Advance planning is great, but advance food preparation must be done with great care.

— Leftovers take extra preparation. There's a temptation to heat leftovers to serving temperatures and serve them. But beware! Leftover, cooked meat should be heated through and through to at least 165°F (about 74°C) to destroy vegetative cells of perfringens and salmonella. Or the meat should be cut up into small pieces and boiled long enough for the interior temperatures to become lethal to micro-organisms. Once reheated, the leftover food should be served while hot and not allowed to remain at incubating temperatures.

Clean Places

A foodservice establishment should be designed and built for public serving of food. That statement has the ring of simplicity, but its implications are numerous. It means, for one thing, that there must be adequate safeguards against insects and animal pests. It also means that layout and facilities should be selected for their easy cleaning properties. Any repairs, alterations or expansions of the business should be done with highly cleanable materials. Porous surfaces and equipment with crevices should be avoided because they are easy to soil and difficult to clean.

The manager obviously has to rely on his staff to keep the facility spick-and-span, and he is more likely to gain their cooperation if he makes clean-up tasks easier and more convenient. He will of course want to provide them with appropriate cleaning agents and tools. He should also make certain that employee restrooms are adequate, and that handwashing facilities are maintained separate from foodwashing stations to avoid cross-contamination.

Just as food sanitation should be a regular part of employee training, so should clean-up techniques. Workers should be told how to go about cleaning and sanitizing, including the materials to be used and the way to protect foodware from contamination after it has been cleaned.

The most successful manager will organize clean-up and maintenance work in accordance with a regular schedule. He will not want cleaning functions to be performed when premium wage rates prevail, but should plan them to fit in with normal work periods.

To Keep Them Clean

The professional foodservice manager is committed to providing wholesome food. He is guided in that objective by a body of laws and ordinances, regulations and codes. It's his responsibility to operate within these requirements, but he needn't be overwhelmed by the weight of official rules and restrictions. The inspectors who represent the regulatory agencies are there to see to it that he obeys the laws, but they're also public servants who can provide technical advice and assistance. Guidance and information are also available from many government sources, as well as from numerous sources within the industry, including food and equipment suppliers and representatives of foodservice associations. Not the least of these are the sponsors of this text — NIFI and the National Sanitation Foundation — plus many others who are cited in the references. In short, keeping a foodservice facility clean is a big job, but you've got lots of expert help.

Managing a Safe Foodservice

The real secret of "managing the microbe" is managing people. Confronted by all the pressures and details of feeding a demanding public, and surrounded by potential sources of contamination, the foodservice manager must recognize a simple fact of his profession: He can't do it all himself. However exceptional a human being, he can't be manager, purchasing agent, chef, waiter, bus boy, and cashier all rolled into one. He must inevitably honor the classic definition of management, which is "getting things done through people." To achieve management goals in food sanitation, the manager should follow three fundamental approaches: 1. Delegation of responsibility. 2. Training. 3. Self-inspection.

Delegation of Responsibility

As the first step in distributing the work load, the manager assigns tasks to others — he delegates the responsibility. For a long time in the foodservice industry, delegation of responsibility was done by the "magic apron" system. The dishwasher became the chef by the simple act of changing aprons . . . and the menu suffered until his abilities caught up with his promotion. But responsible and effective delegation is a much more sophisticated process which involves training and experience on the part of the supervisor or other employee to whom responsibility and work are assigned. It is obvious that the employee should be oriented to policy and procedures before he starts work. If he is to be held responsible for a task, he must know what that task involves.

If delegation were merely a matter of passing out the work, the manager's obligation would end there. In fact, the employee must still be motivated to do the job on his own, conscientiously and with enthusiasm. Without that incentive, the manager may face the unhappy alternative of having to employ constant supervision. To inspire the willing worker, treat him for what he is — a member of the foodservice team. Modern management theory advises you the manager to sit down with your employee and establish objectives which can be measured in specific terms. At the simplest level, that means you and the worker decide, for example, that he is ultimately capable of preparing 200 servings of soup a day. Since he contributed to the decision, he is more inclined to work to achieve that goal. This same process may be applied to any aspect of the foodservice workload, including sanitation operations. For best results, successful performance on the job should be covered by a system of rewards.

Training

Human memory is frail, and under the pressure of peak dining periods or the abrasion of time, some of the manager's instructions on safe foodhandling may be forgotten. The chef may overindulge the desirable but dangerous practice of keeping a dish on low heat to avoid loss of flavor, a bus boy may wash his hands less frequently, a waitress may stop being so careful about where she sneezes. The answer to all these failings is regularly scheduled training sessions. The manager will want to cover current problems, of course, but he'll also want to review job responsibilities and good sanitation practice. Frequent training sessions serve to reinforce memories so that safe procedure becomes a habit.

Self-Inspection

The manager's responsibility for safe foodservice doesn't end when he assigns a trained employee to the job. He must continue to make certain that tasks are being properly performed, through regular, on-the-spot inspections. Every aspect of the operation deserves the manager's personal attention, no matter how reliable his workers may be.

There are so many small but important details that should be reviewed, it would be easy to overlook some. The manager of even a small foodservice may find it desirable to make his inspections on the basis of a written checklist. Model checklists are contained in "A Self-Inspection Program for Foodservice Operators," published by the National Restaurant Association. This publication is designed so that the checklists may be removed and copied for repeated use.

Employees shouldn't be made to feel that they're being "graded" or spied on, but that their work is important, that the boss is interested in what they do, how they do it, is receptive to better ideas and may have a few of his own! The difference here is between a manager demonstrating a genuine interest in operating a thriving business, and an employee thinking, "Look out, the boss is checking up on me again."

In Conclusion

The responsibilities of the foodservice manager are understandably great but certainly not unbearable. After all, he is running one of the most important businesses in the nation — serving the eating-out public — and his products and services are under constant scrutiny.

In studying this book, we trust that the manager and the prospective manager have mastered the precepts and are prepared to practice the principles of safe foodservice management. While an eating establishment may continue in business even if it serves food of mediocre quality, it is unlikely to survive for long with a reputation for unsafe food.

It is hoped that the foregoing pages will help you in pursuing a successful and rewarding career in the foodservice industry.

Good luck and good management!

Appendix A

A SELF-INSPECTION PROGRAM
FOR FOODSERVICE OPERATORS

On Sanitation
and Safe Food Handling

Published in 1973 by the
NATIONAL RESTAURANT ASSOCIATION
as a service of the NRA Public Health
and Safety Committee

Reprinted in abbreviated format by permission of the publisher.

FOOD HANDLING PRACTICES

Date Inspected: _____

Inspected by: _____

ITEM	YES	NO	Comments on Deficiencies Noted and Action Required	Date Corrected
Is food, in pans or containers, on floor?				
Are perishable or potentially hazardous foods being held at room temperature?				
Are fruits and vegetables thoroughly washed prior to preparation and serving?				
Are food warmers, steam tables and bain-maries used to reheat prepared foods?				
Are frozen foods being properly thawed under refrigeration or under cold running water or cooked directly from frozen state?				
Are raw, and cooked or ready to serve foods being prepared on the same cutting board without washing and sanitizing the board between changed use?				
Are hands being used to pick up rolls, bread, butter pats, ice, or other food to be served?				
Are waitresses or busboys handling place settings and serving food without washing hands after wiping tables and bussing soiled dishes?				
Are food servers touching food contact surfaces of plates, tumblers, cups and silverware when setting table or serving customer?				
Is floor being swept while food is exposed, being served or when customers are eating?				

Table of Contents

GUIDELINES

WHY SELF-INSPECTION?

The initial impression a customer receives of your establishment determines to a considerable degree the extent of his future patronage. Very important to you is what the customer sees during his visit . . . hopefully, a bright, clean and attractive place in which to dine. More important to him is what your customer, for the most part, does not see—the personal hygiene practices of your employees; the manner in which food is handled during storage, preparation and serving; and the manner in which equipment and utensils are cleaned and sanitized.

The extent to which your employees practice good personal hygiene, prepare and serve food in a safe manner and maintain equipment and utensils in a clean and sanitary condition, measures the extent to which you protect your customers from contaminated food and possible foodborne illness.

You cannot rely on the visits and counsel of your health department's sanitarian to assure the day to day protection necessary to present a safe eating place for your customers. Although your customers visit your restaurant every day, the sanitarian sees it infrequently . . . once a month; once every 2 or 3 months; once or twice a year or, perhaps, not at all.

As in many other situations, a foodservice operator has to "do it himself." The effective protection of your customer can best be achieved by requiring that your managers and supervisors do a continual job of self-inspection of both facilities and practices . . . by checking department-by-department and function-by-function for unsafe procedures and unsanitary conditions which can lead to contamination of food by harmful bacteria.

WHO SHOULD MAKE THE INSPECTIONS?

The following members of your organization should be designated to perform the inspections of both facilities and activities in your foodservice establishment:

1. Yourself—the owner/operator or manager.
2. Departmental supervisors.
3. The night manager or supervisors.
4. Supervisors of satellite operations.
5. Sanitarian consultants under contract to perform sanitation inspections.

WHAT ACTIVITIES AND FACILITIES SHOULD BE COVERED?

Your sanitation and safe food handling self-inspection program should include coverage of the following:

1. Personal safeness of food handlers.
2. Food handling practices.
3. Raw food and ingredients receiving points.
4. Food storage facilities.
5. Food preparation areas.
6. Food holding equipment.
7. Dining rooms and serving areas.
8. Ware washing and storage facilities.
9. Customer rest-rooms.
10. Employee facilities—toilets, lavatories, locker-rooms, lunch-rooms.
11. Storage facilities for supplies and equipment.
12. Garbage and trash storage and disposal areas.
13. Boiler-rooms, compressor installations and other utilities.
14. Entry ways, exits and outside areas including parking lots, drive-in service marquees, refuse disposal facilities, etc.
15. Vehicles used for transporting food.
16. Aspects of customer concern.

HOW FREQUENTLY SHOULD INSPECTIONS BE ACCOMPLISHED?

Inspection frequencies should be based on the relative extent to which hazards to health are involved. Critical points in operational methods and procedures and in the use of equipment which are directly related to contamination of food and growth of harmful bacteria should be identified. The priority and frequency of inspection checks should then be based on relative importance of functional areas and activities to the likelihood of food contamination.

Your program of self-inspection is not primarily for the purpose of preparing for regulatory agency inspections but, rather, to provide a continual check on the adequacy and completeness of supervision, and on the safeness of all areas and activities of your operations. Consequently, your scheduling of the inspections to be accomplished of specific activities and functions should involve frequencies appropriate to assure you that deficiencies are discovered and corrective steps are taken.

HOW SHOULD INSPECTION CHECK LISTS BE DEVELOPED?

Model sanitation inspection check lists developed by enforcement agencies, public health departments, educational institutions, restaurant associations, manufacturers of foodservice sanitation equipment and supplies and foodservice companies are available in abundance. However, these inspection check lists are not specifically tailored to your particular operation.

It is recommended, therefore, that you seek assistance from your local regulatory agency or sanitarian consultant in accomplishing the necessary inspection sheets for your operations to ensure that:

1. They are specifically applicable to your facilities and company policies, practices and procedures.
2. They are compatible with the food-service establishment sanitation regulations which apply to your operation.
3. They are divided to permit separate inspection checks of specific functional areas or procedures.

The inspection check sheets included in this kit are provided for your use in building a self-inspection program pertinent to your operation. You will find that, when you have discarded those which do not apply to your operation, you will have a set applicable to your needs.

WHAT FOLLOW-UP ACTIONS SHOULD BE TAKEN?

You will want to initiate appropriate action to correct undesirable situations which your self-inspections may reveal and you will, no doubt, include mention of some of them during training sessions for employees or through the media of your employee bulletins, newsletters and posters.

It would be useful for you to maintain a file of your self-inspection reports to permit your management review of (1) the immediate and long-range effectiveness of your self-inspection program, of (2) the effectiveness of supervision over your operations and of (3) the promptness and adequacy of action to correct unsatisfactory procedures and conditions.

WHAT BENEFITS MAY ACCRUE TO THE FOODSERVICE OPERATION?

When a file is maintained on the self-inspections accomplished and on follow-up actions taken to correct undesirable situations and to bring deficient procedures and conditions to the attention of your employees, a number of benefits can result:

1. When copies are furnished to your regulatory department or shown to your regulatory department's sanitation inspector, they will know the extent of your interest in maintaining a safe operation and of your continuing program to identify and correct unsafe conditions. This could lead to a reduction in the number of official inspections made of your establishment, as the regulatory department turns its attention to problem establishments. Regulatory officials will also be able to recognize conditions or situations with which you are having difficulty and, in many cases, will be able to assist you in resolving them.

2. Your self-inspection records will be useful as evidence of your interest and effectiveness in maintaining a safe and sanitary operation in case you are confronted by representatives from the news media or consumer groups.

3. Your continuing review of both inspection results and follow-up action will, undoubtedly, protect you from law suits and claims because of action you have taken to reduce or eliminate illness causes.

PERSONAL SAFENESS
Infections and Illness, hygiene and grooming

Do any food handlers have infected burns, cuts, boils?

Do any food handlers have acute respiratory illness?

Do any food handlers have infections or contagious illness transmittable through foods?

Are food handlers wearing clean outer garments?

Are food handlers free of body odors?

Are food handlers' hands clean—washed at start of work day and as frequently as necessary?

Are food handlers wearing hats, caps or hairnets or other effective hair restraints?

Are food handlers observed picking nose or pimples, scratching head or face?

Are food handlers observed smoking or eating in food preparation or serving areas?

Are fingernails of food handlers short and clean?

Are instances of spitting in sinks, on floor or in disposal area observed?

Are food servers seen to cough in hands?

Are food handlers wearing rings (other than plain band), dangling bracelets, wrist-watches, etc., while preparing or handling food?

Are cooks' wiping cloths used to wipe off perspiration on face used for no additional purpose?

Have all employees been instructed on minimum sanitation and food protection requirements?

FOOD HANDLING PRACTICES

Is food, in pans or containers, on floor?

Are perishable or potentially hazardous foods being held at room temperature?

Are fruits and vegetables thoroughly washed prior to preparation and serving?

Are food warmers, steam tables and bain-maries used to reheat prepared foods?

Are frozen foods being properly thawed under refrigeration or under cold running water or cooked directly from frozen state?

Are raw, and cooked or ready-to-serve foods being prepared on the same cutting board without washing and sanitizing the board between changed use?

Are hands being used to pick up rolls, bread, butter pats, ice, or other food to be served?

Are waitresses or busboys handling place settings and serving food without washing hands after wiping tables and bussing soiled dishes?

Are food servers touching food contact surfaces of plates, tumblers, cups and silverware when setting table or serving customer?

Is floor being swept while food is exposed, being served or when customers are eating?

RAW FOOD AND INGREDIENTS RECEIVING POINT

Is food inspected immediately upon receipt for spoilage or infestation?

Is perishable food moved promptly to refrigeration?

Are unattended perishable food deliveries on loading dock or dolly?

Are non-food supplies checked for infestation?

Are empty shipping containers and packing removed to disposal area promptly?

Is receiving area free of food particles and debris?

Is floor of receiving area clean?

DRY STORES

Is all food stored off the floor—on shelves, racks or platforms?

Is the floor clean and free from spilled food?

Are shelves high enough off floor to permit cleaning underneath, or is area beneath shelf enclosed to preclude accumulation of soil?

Are shelves away from wall to permit ventilation and discourage nesting of insects and rodents?

Have empty cartons and trash been removed?

Are canned goods removed from cartons for shelving to maximum extent practicable?

Are food storage shelves clean and free of dust and debris?

Are food supplies stored in a manner to insure "first-in"—"first-out" use?

Is storeroom dry—free from dampness?

Are non-food supplies stored separately from food stock?

Are all toxic materials, including pesticides, conspicuously labeled and used from original containers only?

Are pesticides separately stored in a well-marked cabinet?

Is there evidence of insects or rodents?

Is there evidence of misuse or spillage of insecticides or rodenticides?

REFRIGERATOR STORAGE

Are all refrigerators operating?

Are refrigerators equipped with accuate themometers?

Are refrigerators maintaining potentially hazardous foods at temperatures of 45°F or lower?

Are refrigerators clean and free from mold and objectionable odors?

Is all potentially hazardous food, not in actual preparation or hot holding, stored under refrigeration?

Is all food being stored off the floor of walk-in refrigerators?

Are foods stored on shelves spaced to provide for adequate air circulation?

Are panned raw or cooked foods, on shelves, covered to prevent contamination?

Are cooked foods such as ground meat, stew, dressing or gravy not stored in large-quantity containers?

Are foods stored in a manner to permit "first-in"—"first-out" use?

Are any spoiled foods present?

Are raw foods stored separately from cooked foods?

Are shelves high enough from the floor to permit cleaning underneath?

Are shelves free from food husks, leaves, wrappings or debris?

Is traffic in and out of walk-in refrigerators excessive?

Are any refrigerators overloaded?

FREEZER STORAGE

Are freezer storage units operating?

Do all boxes or cabinets have accurate thermometers?

Are freezer storage units maintaining an interior temperature of 0° F, or lower?

Is there excessive traffic in and out of walk-in freezer storage boxes?

Is food stored in a manner which permits "first-in"—"first-out" use?

Is food stored in a manner to insure air circulation?

Do cabinet walls or coils need defrosting?

VEGETABLE PREPARATION

Is the vegetable preparation area clean and free from objectionable dampness and odor?

Are non-refrigerated vegetables stored in ventilated bins or in crates on elevated platforms?

Is area free from empty containers and debris?

Is vegetable sink(s) used for hand washing or for warewashing?

Is vegetable sink used for dumping mop water or pan drippings?

Are peelers or paring knives present in vegetable sink?

Are vegetable preparation equipment parts washed in the vegetable sink?

Are vegetable peelers, slicers, choppers, etc., not in use, clean?

Are vegetable peelers, slicers, choppers, etc., being cleaned between changed uses?

MEAT-CUTTING AREA

Is meat-cutting area clean and free from objectionable odor?

Is area free from accumulation of meat cartons, wrappings and other debris?

Are meat-cutting wastes discarded into approved containers and removed to the disposal area?

Are cutting boards in good condition—free from splits, holes or cuts?

Are cutting boards cleaned and sanitized between changed use?

Are all cutting boards, tables, grinders, slicers, meat saws, boning knives and other meat-cutting equipment clean and sanitized, if not in use?

Is raw meat awaiting preparation or processed meat cuts, in containers, off the floor?

Is raw meat awaiting processing or processed cuts being held for excessive periods at room temperature?

Is frozen meat, poultry or fish being thawed out with warm water?

Are floors clean and free from sawdust and accumulation of meat particles and blood?

BAKING AREA

Is bakery area clean?

Is bakery area free from empty cartons, containers and debris?

Is bakery area dry?

Are flour and other non-perishable bakery food ingredients off the floor?

Are dough mixers, kettles and other bakery equipment and utensils which are not in use, clean?

Are mixing bowls, pots, and other bakery utensils stored in a manner to prevent splash and contamination?

Are potentially hazardous bakery ingredients and unbaked fillings and liquid mixes not held at room temperature longer than absolutely necessary during preparation?

Are pie fillings not being held and cooled in stock pots or other large containers and at room temperature?

Are pesticides or cleaning supplies not stored or present in the bakery area?

Is canvas not in use as work surface on baker's table?

FOOD PREPARATION AND HOLDING

Is food preparation area generally clean and free from accumulated debris?

Is floor of kitchen and other food preparation and service areas clean and dry?

Is food preparation equipment not in use, clean?

Are utensils not in use clean, sanitized and stored in a manner which will protect them from contamination?

Is preparation equipment cleaned and sanitized between changed use? (This especially pertains to grinders, slicers, choppers and mixers, and to knives).

Are cleaning supplies and pesticides present in the food preparation and service areas?

Are cooks' sinks being used for employee hand washing or dumping of mop water?

Is unused equipment or utensils stored behind ranges, ovens or in other floor spaces in the kitchen area?

Is there evidence of rodents or insects in the kitchen or serving lines of the establishment?

Are thermostats operating and accurate on ranges and deep fat fryers?

Is hot-holding equipment maintaining food at or above 140°F?

Are cold foods being held at 45°F or lower?

DINING ROOM
AND SERVING AREA

Is dining area, including floor, tables and chairs, clean and dry?

Is tableware clean and sanitized and stored in a manner to prevent splash and contamination?

Are single-service items stored and dispensed in a sanitary manner?

Are single-service items disposed of after single use?

Are clean and sanitary cloths used for wiping dining tables and chairs?

Are silverware and serving utensils stored and presented in a manner to prevent contamination and to insure their being picked up by the handles?

WARE-WASHING
AND STORAGE

Is there sufficient hot water to meet the warewashing requirements of the establishment?

Are wash and rinse temperatures proper for the type of machine, and ware being washed, being maintained? (See Manufacturer's Specifications on data plate.)

Is rinse temperature of at least 170°F being maintained for tableware and utensils? (Manual dishwashing.)

Are dishes and utensils being prescraped and flushed prior to washing?

Is detergent concentration being maintained at the necessary level for effective washing?

Are separate workers assigned to remove and store clean tableware, or do warewashing personnel wash hands between handling soiled tableware and sanitized ware?

Is warewashing equipment cleaned after each day's use to remove chemicals, food particles, soil and debris?

Are jets and nozzles cleaned of food particles and other obstructions and contaminants?

Are cleaned and sanitized wares and utensils stored off the floor and in a clean, dry location?

Is improper toweling of tableware and utensils observed?

EMPLOYEE FACILITIES
Toilets, lavatories, locker rooms, lunchrooms

Are employees' facilities clean, dry and free from odor?

Is there sufficient soap, towels and tissue for employee needs?

Is all sanitary equipment operational and in good repair?

Are proper receptacles available for waste materials?

Are these receptacles emptied frequently?

Are soiled cook's aprons, whites and other soiled clothing improperly stored in lockers or left in the facilities?

Is unwrapped food stored in lockers or left in employee facilities?

Is there evidence of rodents or insects in the facilities?

Are containers provided for soiled cooks' whites and employees' uniforms?

STORAGE ROOMS FOR
SUPPLIES AND EQUIPMENT

Are storage facilities for supplies and equipment clean, dry and free of trash and debris?

Are storage facilities free of empty cartons and wrappings which might provide nesting for rodents?

Are supplies stored in a neat and orderly manner?

Are supplies stored off the floor and away from walls to permit access for cleaning and to prevent harborage of rodents and roaches?

Is perishable or unpackaged food present?

Are containers of pesticides not in marked cabinet present?

Is there evidence of rodents or insects?

GARBAGE/TRASH STORAGE AND DISPOSAL AREAS
Including incinerators, compactors, etc.

Is area generally clean and orderly?

Is floor, platform or ground surface free from spilled particles of food?

Is area free from objectionable odor?

Is grease and other liquid spillage on floor, platform or ground surface?

Are spilled food particles and litter present in front of incinerator, dumpsters, etc.?

Are trash and garbage containers clean on the outside?

If can liners are not in use, are all garbage containers closed with tightfitting covers?

Is trash confined in orderly fashion or in suitable containers?

Is there an accumulation of trash because of infrequent pick-up?

Has trash accumulated in corners, under stairs or platforms?

Are there puddles of wash water and food particles and liquids?

Is there any evidence of rats, ratholes or nests in the vicinity of the disposal area?

Are empty garbage and refuse containers washed prior to being returned for use?

Is mop water properly disposed of?

BOILER ROOMS, COMPRESSOR AND OTHER UTILITY AREAS

Are boiler rooms, compressor rooms and other utilities rooms clean, dry and free of foods, soiled or greasy utensils and food preparation equipment?

Are they free of soiled linen and rags, empty containers and cartons, trash and debris?

Is there evidence of rodents and insects?

ENTRYWAYS, EXITS
AND EXTERIOR AREAS
Including parking lots and driveways

Are entryways clear of trash and debris?

Are doors and screen doors tight fitting to prevent entry of insects and rodents?

Is there any evidence of rat holes or entry points near or into the building?

Are there wet spots or pools, or high grass or weeds which could form breeding spots for insects?

Is the rear area, parking lot or surrounding area free of litter, trash and debris?

Do noxious birds nest or roost in ledges or eaves of the establishment?

VEHICLES USED FOR
TRANSPORTING FOOD

Is the cargo area of the vehicle thoroughly clean and free from dirt and debris?

Has all food in containers been removed for proper disposal or storage?

Is potentially hazardous food being carried at proper temperatures of heat or cold?

Is food being carried in adequately insulated containers?

Are all food spills on shelving or floor washed from the vehicle after each use?

Is there any evidence of insect infestation of the vehicle body?

RESTROOMS

• CUSTOMER CONCERNS

Are customer restrooms clean, dry, light, well ventilated and free of odor?

Is all sanitary equipment operating satisfactorily?

Is there an adequate supply of soap, towels and tissue?

Are waste containers covered and kept clean?

Are waste containers emptied frequently?

Is there adequate hot and cold water?

Are all drains operating in a satisfactory manner?

Is there any sign of rodents or insects?

Are toilet doors self-closing and in good working order?

ENTRANCE AND FOYER

• CUSTOMER CONCERNS

Is the entryway and waiting room clean and attractive?

Is it free from litter?

Are chairs and benches clean and lamps and fixtures clean and free from dust?

Are posters and printed materials clean and neatly racked or posted?

Does the customer's first view of your establishment convey the image of cleanliness and freshness?

GENERAL CLEANLINESS
OF DINING AREA

• CUSTOMER CONCERNS

Is the floor dirty, dingy or littered, particularly with food particles and napkins?

Are tables streaked and condiments dirty?

Are there crumbs, spilled liquid on chairs or benches?

Are menus food-marked or worn and dirty?

Are table linens food-marked? Are they tattered or torn?

Is tableware cracked, chipped, streaked or food-soiled?

Is silverware thumb-marked, dingy, spotted or food-soiled?

Are soiled dish trays left near customer tables?

A point of safety—are insect sprays being used when food is exposed or customers are present?

CLEANLINESS OF
SERVICE PERSONNEL

• CUSTOMER CONCERNS

Are waitress uniforms crumpled or soiled?

Are waitresses using strong or offensive perfume?

Are servers sniffing, coughing or rubbing or wiping nose?

Do servers handle drinking glasses by their tops or silverware by their blades, tines or bowls?

Are waitresses wearing shaggy hair-dos and wigs?

Do cooks and servers smoke in view of customers?

Do servers handle rolls, butter, ice, etc., by hand in filling dishes and water glasses?

Do cooks, servers or busboys scratch head, face or body in view of customers?

Does server touch food with thumb or fingers when serving plated food?

SENSORY FACTORS

• CUSTOMER CONCERNS

1. Heat

Is dining area too hot or cold for customer comfort?

Is heat, or steam, from serving line unpleasant for customers (and the servers)?

2. Light

Is light in dining area too bright and glary?

Is light too dim, making it difficult for customer to see the menu and his food and tableware?

3. Noise

Is clatter of dish- and warewashing offensive to the customer?

Are busboys too noisy in handling removal of soiled tableware?

Is loudness of waitresses, cooks or busboys offensive and distracting to the customer?

4. Odors

Do kitchen odors greet the customer as he enters the dining room?

Is there an "old grease" odor in the dining room or as exhausted to the street or parking area?

Is the odor of strongly-flavored food allowed to dominate the establishment?

Is spoiled food disposed of promptly to prevent obnoxious odor?

Appendix B

BIBLIOGRAPHY

(1) *Control of Communicable Diseases in Man,* 10th ed. John E. Gordon, Editor. American Public Health Association, 1790 Broadway, New York, N.Y., 1965.

> A 282-page APHA report designed as a text for public health workers and as a guide for public health administrators. Contains source materials for use in drafting health regulations for the control of communicable disease, and in the development of educational programs in the field of public health.

(2) *Food Sanitation.* Rufus K. Guthrie. AVI Publishing Co., Inc., Westport, Conn., 1972.

> A reference text of 247 pages presenting an overview of food sanitation, covering common diseases and their modes of transmission. Representative sanitation laws and ordinances and recommended health codes are included. Written to give the food industry worker and layman an understanding of the biological principles involved in air, land and water pollution control. A commonsense book on food protection and public health.

(3) *Food Poisoning and Food Hygiene,* 2nd ed. Betty C. Hobbs. Edward Arnold, Ltd., London, 1968.

> A 252-page reference work developed for teachers and advanced students of food poisoning prevention and control. Fifteen chapters and appendices provide suggested lecture material on sanitation in the kitchen and the food shop. Recommended for health officers, canteen supervisors, food store managers, and instructors in catering and domestic science.

(4) *Quantity Food Sanitation*. Karla Longrée. Interscience Publishers, a Division of John Wiley and Sons, Inc., New York, 1968.

> A comprehensive one-volume work on foodborne illnesses, their causes and control (397 pages). For many purposes the definitive text on sanitation in quantity foodservice. Microbiological aspects treated in some depth.

(5) *Sanitary Techniques in Food Service*. Karla Longrée and Gertrude G. Blaker. John Wiley and Sons, Inc., New York, 1971.

> A reference text of 225 pages providing practical guidance in culinary sanitation for the foodhandler. Designed as a teaching aid for vocational training, the book contains many charts and tables depicting safe operating procedures.

(6) *Microbiology for Sanitary Engineers*. Ross E. McKinney. McGraw-Hill Book Company, Inc., New York, 1962.

> A reference text, one of a series designed primarily for the practicing and student sanitary engineer. Written by an engineer who became a microbiologist. Oriented to technology of waste treatment but general principles apply to food microbiology.

(7) *A Self-Inspection Program for Foodservice Operators*. National Restaurant Association, One I.B.M. Plaza, Chicago, Ill. 60611, 1973.

> Guidelines for a sanitation and safe food-handling self-inspection program. Contains a complete set of check-sheets covering the various areas of a commercial foodservice operation.

(8) *Career Ladders in the Foodservice Industry*. National Restaurant Association, One I.B.M. Plaza, Chicago, Ill. 60611, 1971.

> A study report pointing out important elements and deficiencies of in-service training programs in foodservice establishments.

(9) *Profits and Your People.* National Restaurant Association, One I.B.M. Plaza, Chicago, Ill. 60611, 1972.

> A training guide providing teaching techniques and methods in the areas of foodservice sanitation and customer interests.

(10) "The Sub-Standard Washroom." National Restaurant Association, One I.B.M. Plaza, Chicago, Ill. 60611. (*Food Service Research Digest,* Autumn 1966).

> A brief survey of items which impress the customer favorably and unfavorably in the washrooms of commercial eating establishments.

(11) "Know Your Health Officer." National Restaurant Association, One I.B.M. Plaza, Chicago, Ill. 60611, 1962.

> A bulletin giving the foodservice operator basic information on the background, organization and operation of public health regulatory agencies.

(12) *48 Ways to Foil Food Infections.* Channing L. Bete Co., Inc., Greenfield, Mass., 1970 edition.

> A 15-page illustrated pamphlet describing, in layman's terms, the major bacterial foodborne illnesses, their causative agents, modes of transmission and control.

(13) "Attitudes About Sanitation in Restaurants." National Restaurant Association, One I.B.M. Plaza, Chicago, Ill. 60611 (*Food Service Research Digest,* Autumn 1966).

> A tabulation of survey data and descriptions of survey findings giving the public's reaction to sanitation and other problems in the foodservice industry.

(14) "Sanitation Checklists for Management." (*Cooking for Profit,* March 1972).

> Procedures and checklists helpful to management in preparing and implementing cleaning schedules.

(15) *Hot Tips on Food Protection.* Pub. No. 1404. Public Health Service, U.S. Department of Health, Education, and Welfare.

> An illustrated pamphlet providing guidance on the hot-holding of food.

(16) *Pest Prevention.* National Restaurant Association, One I.B.M. Plaza, Chicago, Ill. 60611.

> An informative bulletin for the foodservice operator and the professional pest control operator on prevention and control of common insect, rodent and bird pests.

(17) *Food-Borne Illnesses (Reference chart).* National Restaurant Association, One I.B.M. Plaza, Chicago, Ill. 60611.

> A foldout chart providing information on leading food-borne diseases. Tabulates data on the causative agent, foods usually involved, how food is contaminated, and preventive and corrective procedures for a list of four high-incidence foodborne ailments and 11 of less frequent occurrence.

(18) "What You—a Food Service Operator—Should Know About Salmonellae" (from remarks by Dr. E. M. Foster). National Restaurant Association, One I.B.M. Plaza, Chicago, Ill. 60611 (*Food Service Research Digest,* Summer 1967).

> A four-page summation presenting general information on salmonellosis, with data on outbreak sources, control measures applied, and an answer to the question, "What can the food service operator do?"

(19) *Discover the Unseen World, Prevent Food Poisoning.* Cooperative Extension Service, Michigan State University, Extension Bulletin No. 411, Home and Family Series, May 1966.

> A 13-page bulletin of general information on bacterial illnesses transmitted by food. Document is oriented toward domestic use, but language and content also lend themselves to commercial applications.

(20) *Manual on Sanitation Aspects of Food Service Equipment.* National Sanitation Foundation, Ann Arbor, Michigan 48105, approved by the NSF Advisory Committee on Installation of Food Equipment, 1968.

> A guide for the sanitary installation of foodservice equipment.

(21) *Standards on Food Service Equipment.* National Sanitation Foundation, Ann Arbor, Michigan 48105.

>A description of the fundamentals of foodservice equipment design and construction, and the materials used therein. Evaluation procedures used by the NSF Testing Laboratory are described.

(22) *Sanitary Design and Evaluation of Food Service Equipment.* C. A. Farish. National Sanitation Foundation, Ann Arbor, Michigan, 1971.

>A summary of the importance of good design and construction features, and materials selection in the fabrication of safe foodservice equipment. Operation of the NSF Testing Laboratory also covered.

(23) *Food Service Sanitation Manual.* Pub. No. 934 (1962) Public Health Service, U.S. Department of Health, Education, and Welfare.

>Official federal guidelines containing 1962 Public Health Service recommendations for "A Model Food Service Sanitation Ordinance and Code" presented for adoption by state and local jurisdictions or incorporation into their laws and regulations. Annotated with descriptive matter on reasons for the code's provisions and on compliance criteria.

(24) *Current Concepts in Food Protection.* Food and Drug Administration Course Manual, 1973. Public Health Service, U.S. Department of Health, Education, and Welfare.

>A training manual used in seminars for field sanitarians involved with food protection.

(25) "Protecting our Food," *The Yearbook of Agriculture,* 1966. Jack Hayes, Editor. U.S. Department of Agriculture, Government Printing Office, Washington, D.C. 20402.

>A comprehensive reference explaining how food protection problems were solved in the past, and how new developments will help solve future problems. A practical exposition of the proper way to handle foodstuffs from farm to table; marketing trends; and an overview of government regulation and administration.

(26) "Emerging Foodborne Diseases." Frank L. Bryan, Ph.D. *Journal of Milk and Food Technology*, Part I, Oct. 1972, Vol. 35, No. 10; Part II, Nov. 1972, Vol. 35, No. 11.

A technical summation of the impact of changes in food production, processing and preparation. Includes results of a survey disclosing factors contributing to 493 foodborne disease outbreaks during a 10-year period (1960-1970).

(27) *Foodborne Diseases of Contemporary Importance.* Frank L. Bryan, Ph.D. Center for Disease Control, Atlanta, Ga. 30333. Public Health Service, U.S. Department of Health, Education, and Welfare.

A 35-page compilation of technical publications relating to the more common foodborne illnesses. Provides detailed information on disease incidence and prevention.

(28) *Environmental Health and Safety in Health-Care Facilities.* Richard C. Bond, George S. Michaelsen and Roger L. DeRoss. Macmillan Publishing Co., Inc., New York, 1973.

A comprehensive reference on health-care facilities. Chapters on food sanitation are appropriate for general commercial operations.

(29) *Foodborne Outbreaks Annual Summary 1971.* Center for Disease Control, Atlanta, Ga. 30333. Public Health Service, U.S. Department of Health, Education, and Welfare.

An official document of the Center for Disease Control tabulating reported foodborne outbreaks in the United States for the year.

(30) *Salmonellosis.* Extension Bulletin No. 339. Edmund A. Zottola. Agricultural Extension Service, University of Minnesota, 1967.

A 14-page illustrated bulletin providing pertinent information on salmonellosis. Useful to the domestic and commercial foodhandler.

(31) *Staphylococcus Food Poisoning.* Extension Bulletin No. 354. Edmund A. Zottola. Agricultural Extension Service, University of Minnesota, 1968.

> An illustrated bulletin providing general and specific information on staphylococcus food poisoning. An excellent domestic or commercial guide.

(32) *Clostridium Perfringens Food Poisoning.* Extension Bulletin No. 365. Edmund A. Zottola. Agricultural Ex-Extension Service, University of Minnesota, 1971.

> Descriptions of *C. perfringens* outbreaks.

(33) *Food Service Operations.* NAVSUP Pub. No. 421. Naval Supply Systems Command, 1971.

> An operating manual designed for the general mess in ships and shore stations of the U.S. Navy. Principles and practices described find application in any large, highly organized foodservice.

(34) *Code of Recommended Practices for the Handling of Frozen Food.* Frozen Food Coordinating Committee, 919 18th Street N.W., Washington, D.C. 20006, revised 1970.

> A set of specific recommendations on the handling of frozen foods, advocating voluntary action by industry in preference to compulsory regulations.

(35) *The Preparation of Occupational Instructors.* Office of Education, U.S. Department of Health, Education, and Welfare, Superintendent of Documents No. FS 5:280:80042, 1966.

> A manual on the training of instructors in occupational fields.

(36) *Programmed Cleaning and Environmental Sanitation for Buildings, Plants, Offices and Institutions.* John C. Gardner, Editor. The Soap and Detergent Association, New York, 1971.

> A compendium of management procedures and methods covering general and specific aspects of housekeeping and sanitation.

(37) *Preparing Instructional Objectives.* **Robert F. Mager, Ph.D. Fearon Publishers, Belmont, Calif. 1964.**

> A must for every trainer and a cornerstone book in current educational technology. In easily understood and interesting language, demonstrates how to *define* teaching objectives, how to *state* them clearly and how to describe criteria by which to measure success.

(38) *Sanitary Food Service Instructor's Guide.* **U.S. Department of Health, Education and Welfare, 1969.**

> Detailed guide on teaching the principles of foodservice sanitation.

(39) **"Salmonellosis Among Restaurant Patrons: The Incisive Role of a Meat Slicer."** *American Journal of Public Health,* **Nov. 1973, Vol. 63, No. 11.**

> Background and description of the investigation of a salmonellosis outbreak, with a detailed analysis of results of the investigation.

(40) *Food Service Sanitation.* **Bertha Yanis Litsky. Modern Hospital Press, McGraw-Hill Publications, Chicago, Ill., 1973.**

> A text and sourcebook of 230 pages documenting the worldwide need for improvement in foodservice sanitation based on numerous kitchen surveys conducted by the author. The book espouses a three-way approach to the problem—by the foodservice manager, the sanitarian and the microbiologist. Presents a wealth of technical information on bacteriological sampling and test procedures.

Appendix C

INDEX & GLOSSARY

A

ADULTERANT — *A substance or agent which renders impure, contaminates or debases* — 24.

AEROBIC — *Living or active only in the presence of air.*

AEROSOL — *Material dispensed as a gas or fine mist.*

ALIMENT — *Nourishment, food* — 20.

ALIMENTARY CANAL — *Digestive tract* — 20.

ALKALINITY — *Degree of base (e.g., potassium or sodium carbonate) versus acid content* — 110, 117. SEE ALSO pH.

American Public Health Association, 170

AMINO ACID — *Any of numerous organic compounds, the chief components of proteins* — 25.

AMOEBA — *Any of a large class of unstructured, jelly-like protozoans.*

AMOEBIC DYSENTERY — *Chronic intestinal infection caused by the protozoan* ENTAMOEBA HISTOLYTICA.

ANTIBODY — *A substance that counteracts disease agents in the blood or tissue of an organism* — 22.

ANTICOAGULANT — *A substance that hinders the clotting of blood* — 136.

Association of Food and Drug Officials of the United States, 171.

Association of State and Territorial Directors of Local Health Services, 170

Association of State and Territorial Health Officers, 170

B

BACILLARY DYSENTERY — *Acute intestinal infection caused by rod-shaped bacteria such as* SHIGELLA SONNEI — 46, 47. SEE ALSO SHIGELLOSIS.

BACILLI *(sing. BACILLUS)* — *Rod-shaped bacteria. Bacillus is also a specific genus of bacteria of which the species* BACILLUS CEREUS *is notable as a possible disease agent in starchy foods* — 33.

BACTERIA *(sing. BACTERIUM)* — *A large class of micro-organisms, typically one-celled, many of which are active in the fermentation process whereby organic matter is converted into nutrients for plants; principal agents of foodborne disease* — 20, 22, 30-37. SEE ALSO MICRO-ORGANISM.

    cold-loving, 73

    colonies of, 33

    contamination by, 20

    favorable conditions for growth of, 26, 35

    incubation temperatures, 26

    limited effects of cooking in destroying, 90

    multiplication of, 21, 22, 33-35

    need for prevention of, 21, 22

    pathogenic, 42

    size, shape, and distribution of, 32-34

Bactericides, 112

C

D

E

F

G

H

I

L

LEAVEN, LEAVENING — *A substance or organism (as yeast) which produces fermentation; a ferment* — 35.
Leftover food
    safe holding temperatures for, 89-91
    storage of, 89-91
Lighting, shielding of, 103

M

MACRO-ORGANISM — *Visible form of life.*
MALODOROUS — *Foul smelling.*
Management
    responsibilities of, 124-127, 169
    self-inspection by, App. A
Manager, Foodservice, 59, 60, 63-66
Materials, *see* Wall and ceiling materials
    for floors, 98, 99
Meat, *see also* Beef, Hamburger, Pork, Poultry, Veal, Sausage
    defrosting of, 84, 85
    inspection on receipt of, 76
    storage of, 77
Meat animals, contamination of, 76, 77
Meat and poultry, USDA inspection of, 76
MEDIUM *(pl. MEDIA)* — *In biology, a nutritive substance that supports bacteria, fungi, etc.*
Metabolic changes, 25
METABOLISM — *Processes of chemical change which provide vital energy in living cells.*
Metals as poisons, 48, 49, 51
MICROBE — *Microscopic life* — SEE MICRO-ORGANISM.
MICRON — *One millionth of a meter* — 32.
MICRO-ORGANISM (S) — *Form of life so small as to be visible only with the aid of a microscope* — 21, 26, 30, 37.
    bacteria, 32-35
    distribution, 57
    helpful or harmful, 31
    molds, 36
    pathogenic, 21
    viruses, 36, 37
    yeasts, 35, 36
Milk, *see also* Dairy products
    pasteurization, 78
    storage of, 78
MOLD — *A fungus organism living on decayed organic matter; the wooly growth it causes, as on bread, cheese, etc.* — 36.
MORBIDITY — *The state of being diseased; in foodborne illness statistics, the relative incidence of disease* — 40.

N

O

P

PYRETHRUM — *Immobilizing agent in contact sprays used for insect control; low toxicity contributes to its safe usage in food establishments* — 135.

Q

QUATS — *Quaternary ammonium, chemical compound used in sanitization of equipment and utensils* — 113.

R

RADIOLOGICAL HAZARD — *Possible danger from radioactivity, as to food preserved by irradiation* — 52.
Rats and mice, *see also* Pest control
    extermination of, 131, 132
    prevention of, 131, 132
Refrigeration, effect on enzyme action, 87
Regulation of foodservices, 21, 166-170
Regulatory agencies, *see* Government control
REPLICATION — *The action of reproduction, as when a microbial cell duplicates itself by splitting in two.*
Restaurant, *see* FOODSERVICE
Roaches, *see* Cockroaches
Rodenticides, *see* Pesticides
Rodents, *see* Rats and mice

S

Safe holding temperatures, 88, 89
Salad dressing as a contamination agent, 87, 88
SALMONELLA — *One of a large genus of bacteria, several of which cause food-borne illness* — 45, 47, 57
Salmonella typhosa, 46, 47
SALMONELLOSIS — *Disease caused by a salmonella organism* — 46, 47, 176.
    SEE ALSO SALMONELLA.
SANITARIAN — *A specialist in the science of public health* — 19, 20, 170, 171.
SANITARY — *Safe for health, wholesome* — 23, 24, 108.
SANITIZATION — *Rendering an article or surface safe for health by reducing the microbial count to a harmless level* — 112-119.
Sanitization procedures
    chemical sanitizing, 112-119
    clean-in-place sanitizing, 113
    heat sanitizing, 112
    immersion sanitizing, 113
    power-spray sanitizing, 113
SAPONIFY — *To convert fat to soap.*
Sausage

T

U

Uncooked foods, contamination danger in, 87, 88
U.S. Center for Disease Control, 43, 166
U.S. Department of Agriculture (USDA), 70, 166
U.S. Department of Health, Education and Welfare (HEW), 70
U.S. Department of the Interior, 70, *see also* Fish; Shellfish.
U.S. Food and Drug Administration (FDA), 7, 15, 16, 166, *see also* Government
    control
U.S. Environmental Protection Agency, 166
U.S. Public Health Service, 23, 70, 166, 169
Utilities, design of, 102

V

Veal
    contamination of, 77
    storage of, 77
Vegetables, *see* Fruits and vegetables, fresh
VEGETATIVE CELL — *Non-sporulating bacterium, or a spore-former in the
    growing, pre-sporulation stage* — 34.
Ventilation, 103
VERMIN — *Any harmful or pestiferous animal, usually insects or small mam-
    mals, as flies, roaches and mice, which infest food and food establishments*
    — 72, 76.
VIRULENT — *Extremely poisonous or harmful; deadly.*
VIRUS — *In microbiology, an extremely small (sub-microscopic) organism which
    subsists only in living media* — 22, 36, 37.
VITAMIN — *A constituent of food vital to nutrition* — 22.

W

Walls and ceilings
    cleaning of, 124
    materials for, 99
Ware-washing machines, *see* Dishwashing machines
Work tables, cleaning of, 124

Y & Z

YEAST — *A subdivision of the fungi group; agent in the fermentation process
    which converts sugar to alcohol and carbon dioxide; leaven; ferment* — 22,
    35, 36.
Zinc poisoning, 48, 49